P9-DVG-941

Algorithms On Graphs

H.T. Lau

TAB Professional and Reference Books

Division of TAB BOOKS Inc.

Blue Ridge Summit, PA

A Petrocelli book

FIRST EDITION
FIRST PRINTING

Library of Congress Cataloging in Publication Data

Lau, H. T. (Hang Tong), 1952—
Algorithms on graphs / by H.T. Lau.
p. cm.

ISBN 0-8306-3429-0
1. Graph theory—Data processing. I. Title.
QA166.L38 1989
511′.5—dc20 89-36595
CIP

TAB BOOKS Inc. offers software for sale. For information and a catalog, please contact TAB Software Department, Blue Ridge Summit, PA 17294-0850.

Questions regarding the content of this book should be addressed to:

Reader Inquiry Branch
TAB BOOKS Inc.
Blue Ridge Summit, PA 17294-0214

CONTENTS

NOTE TO THE READER

Standard graph-theoretic terminology can be found in texts such as F. Harary, *Graph Theory*, Addison-Wesley Publishing Company, 1969; and J.A. Bondy and U.S.R. Murty, *Graph Theory with Applications*, Macmillan Press Ltd., 1976.

The background on graph algorithms and applications can be supplemented by books such as:

A.V. Aho, J.E. Hopcroft and J. D. Ullman, *The Design and Analysis of Computer Algorithms*, Addison-Wesley Publishing Company, 1974.

V. Chachra, P.M. Ghare and J.M. Moore, *Applications of Graph Theory Algorithms*, Elsevier North-Holland, Inc., 1979.

N. Christofides, *Graph Theory, An Algorithmic Approach*, Academic Press, London, 1975.

S. Even, *Graph Algorithms*, Computer Science Press, 1979.

A. Gibbons, *Algorithmic Graph Theory*, Cambridge University Press, 1985.

M. Gondran and M. Minoux, *Graphs and Algorithms*, Wiley-Interscience, 1984.

K. Mehlhorn, *Graph Algorithms and NP-Completeness*, Springer-Verlag, Inc., 1984.

E. Minieka, *Optimization Algorithms for Networks and Graphs*, Marcel Dekker, Inc., 1978.

INTRODUCTION

For convenience, the definitions of most graph-theoretic terms in this book appear in Appendix I. Every chapter is self-contained and largely independent. Each topic is presented in the same format under five subheadings:

A. *Problem description*—a general description of the problem.

B. *Method*—an outline of the solution procedure.

C. *Subroutine parameters*—a description of all parameters of the subroutine that implements the method described in B.

D. *Test example*—a simple example illustrating the usage of the subroutine.

E. *Program listing*—the complete listing of the code.

In general, the solution procedures will be only briefly outlined. References given at the end of this book should be consulted for all details.

Throughout this book, it is assumed that a graph of n nodes and m edges has its nodes numbered from 1 to n. In the implementation of each solution procedure, one of two graph representations is used: the matrix form or the forward star form. The square matrix representation is mainly used to store the edge distance for every pair of nodes in a complete graph, resulting in an n^2 storage requirement. The forward star representation lists each edge by its starting node, ending node, and its length. Furthermore, the edges in the graph are ordered by the starting node so that all edges starting at the same node appear together, resulting in only an $n + 2m$ storage requirement. In this way, if one knows which is the first edge starting at each node i, then one can determine the last edge starting from node i as the edge immediately preceding the first edge starting at node $i + 1$.

A list of all subroutines in this book is summarized in Appendix II. The programs are written in FORTRAN 77. Communication to each subroutine is made solely through the parameter list. The test runs were all performed on the Amdahl 5870 using the IBM VS FORTRAN Compiler.

PREFACE

The many applications of graph theory constantly draw the attention of researchers, especially in the search for efficient algorithms. Although many well-developed procedures have appeared in books and journals, ready-to-use computer codes are generally not easily accessible. This book attempts to provide such a source. It is not meant to be a collection of the most efficient algorithms; the choice of the topics and their solution procedures is purely based on the author's interests. The main objective of this book is to provide computer programs that can be used with minimal effort for problem-solving without much concern for their underlying methodology and implementation.

H.T. Lau
Ile des Soeurs
Quebec, Canada

1

CONNECTIVITY

Maximum Connectivity

A. Problem description

Let n and k be two given positive integers. The problem is to construct a k-connected graph $G(k, n)$ on n nodes with as few edges as possible. Observe that for $k = 1$, the graph $G(1, n)$ is a spanning tree. Consequently, it is assumed that $k \geq 2$. Moreover, it is known that $G(k, n)$ has exactly $\lceil (n*k)/2 \rceil$ edges, where $\lceil x \rceil$ is the smallest integer greater than or equal to x.

B. Method

Label the nodes of the graph by the integers $0, 1, 2, \ldots, n - 1$.

CASE 1. k is even. Let $k = 2t$.

The graph $G(2t, n)$ is constructed as follows. First, draw an n-gon, that is, add the edges

$$(0, 1), (1, 2), (2, 3), \ldots, (n - 2, n - 1), (n - 1, 0),$$

then join nodes i and j if and only if

$$|i - j| \equiv p \pmod{n}, \text{ where } 2 \leq p \leq t.$$

CASE 2. k is odd, n is even. Let $k = 2t + 1$.

The graph $G(2t + 1, n)$ is constructed by first drawing $G(2t, n)$, and then joining node i to node

$$i + (n/2), \text{ for } 0 \leq i < n/2.$$

CASE 3. k is odd, n is odd. Let $k = 2t + 1$.

The graph $G(2t + 1, n)$ is constructed by first drawing $G(2t, n)$, and then join

node 0 to node $(n - 1)/2$,
node 0 to node $(n + 1)/2$,
node i to node $i + (n + 1)/2$, for $1 \leq i < (n - 1)/2$.

C. Subroutine MAKEG parameters

Input:

N	Number of nodes.
K	The required graph is K-connected, K ≥ 2.
NK2	The smallest integer greater than or equal to (N*K)/2.

Output:

INODE, JNODE	INODE(i), JNODE(i) are the end nodes of the ith edge in the K-connected graph, $i = 1, 2, \ldots,$ NK2.

D. Test example

Construct a 5-connected graph on eight nodes with as few edges as possible.

E. Program listing

MAIN PROGRAM

```
      INTEGER INODE(20),JNODE(20)
      N = 8
      K = 5
      NK2 = 20
      CALL MAKEG (N,K,NK2,INODE,JNODE)
      WRITE(*,10) N, K, NK2
 10   FORMAT(/' NUMBER OF NODES =',I3,',',
     +         3X,I2,'-CONNECTED,'/
     +       ' NUMBER OF EDGES =',I3//
     +         ' LIST OF EDGES:'/)
      WRITE(*,20) (INODE(I),I=1,NK2)
 20   FORMAT(1X,25I3)
      WRITE(*,20) (JNODE(I),I=1,NK2)
      STOP
      END
```

OUTPUT RESULTS

```
 NUMBER OF NODES =   8,   5-CONNECTED,
 NUMBER OF EDGES = 20
 LIST OF EDGES:
 1 2 3 4 5 6 7 8 1 1 2 2 3 4 5 6 1 2 3 4
 2 3 4 5 6 7 8 1 3 7 4 8 5 6 7 8 5 6 7 8
```

```
      SUBROUTINE MAKEG (N,K,NK2,INODE,
     +  JNODE)
C
C     Construct a K-connected graph of N nodes with
C       the least number of edges
C
      INTEGER INODE(NK2),JNODE(NK2)
      LOGICAL EVENK,EVENN,JOIN
C
C     Make an N-gon
C
      NK2 = 0
```

```
      N1 = N - 1
      DO 10 I = 1, N1
        NK2 = NK2 + 1
        INODE(NK2) = I
        JNODE(NK2) = I + 1
 10   CONTINUE
      NK2 = NK2 + 1
      INODE(NK2) = N
      JNODE(NK2) = 1
      IF (K .EQ. 2) RETURN
C
      EVENK = .TRUE.
      KHALF = K / 2
      IF (K .NE. 2*KHALF) EVENK = .FALSE.
C
      DO 40 I = 1, N1
        I1 = I + 1
        DO 30 J = I1, N
          JOIN = .FALSE.
          JI = J - I
          DO 20 L = 2, KHALF
            IF ((MOD(L,N) .EQ. JI) .OR.
     +          (JI + L .EQ. N)) JOIN = .TRUE.
 20       CONTINUE
          IF (JOIN) THEN
            NK2 = NK2 + 1
            INODE(NK2) = I
            JNODE(NK2) = J
          ENDIF
 30     CONTINUE
 40   CONTINUE
C
C     If K is even then finish
C
      IF (EVENK) RETURN
C
      EVENN = .TRUE.
      NHALF = N / 2
      IF (N .NE. 2*NHALF) EVENN = .FALSE.
```

```
C
      IF (EVENN) THEN
C
C        K is odd, N is even
C
         DO 50 I = 1, NHALF
            NK2 = NK2 + 1
            INODE(NK2) = I
            JNODE(NK2) = I + NHALF
   50    CONTINUE
      ELSE
C
C        K is odd, N is odd
C
         NPP = (N + 1) / 2
         NMM = (N - 1) / 2
         DO 60 I = 2, NMM
            NK2 = NK2 + 1
            INODE(NK2) = I
            JNODE(NK2) = I + NPP
   60    CONTINUE
         NK2 = NK2 + 1
         INODE(NK2) = 1
         JNODE(NK2) = NMM + 1
         NK2 = NK2 + 1
         INODE(NK2) = 1
         JNODE(NK2) = NPP + 1
      ENDIF
C
      RETURN
      END
```

1-2 Edge-Connectivity

A. Problem description

The problem is to find the edge-connectivity of a given connected undirected graph.

B. Method

As a preliminary, a *network* is defined to be a directed graph G in which each edge (i, j) is associated with a nonnegative number $c(i, j)$ called the *capacity* of the edge. Let the number $f(i, j)$ be the *flow* from node i to node j. A flow in the network is *feasible* if $f(i, j)$ does not exceed $c(i, j)$ for each edge (i, j) in G, and the sum of all flows incoming to node i is equal to the sum of all flows outgoing from node j.

Let s and t be some specified nodes, called the *source* and *sink*, respectively. The *maximum network flow problem* is to find a flow in the network from s to t such that the amount of the flow into t is maximum.

A *cut* is a subset S of the nodes of G with the capacity equal to:

$$\sum_{\substack{i \in S \\ j \notin S}} c(i, j)$$

The well-known max-flow min-cut theorem states that the maximum flow is equal to the minimal cut in a network. The subroutine NFLOW below finds a maximum flow and a minimal cut set in a given network with specified source and sink nodes.

With the background of maximum network flow, the method of finding the edge-connectivity of an undirected graph is quite straightforward.

Denote the nodes of the input connected, undirected graph G by $1, 2, \ldots, n$. For $j = 2$ to n do the following: Take node 1 as the source, node j as the sink in G, assign a unit capacity to all edges in both directions, and find the value of a maximum flow $g(j)$ in the resulting network. The edge-connectivity is equal to the minimum of all $g(j)$, for $j = 2, 3, \ldots, n$.

The subroutine EDGECN below finds the edge-connectivity of a given undirected graph with the help of subroutine NFLOW.

The maximum network flow algorithm requires $O(n^3)$

operations. The edge-connectivity of a graph will therefore be found in $O(n^4)$ operations.

C. Subroutine EDGECN parameters

Input:

N	Number of nodes.
M	Number of edges.
M4	Equal to 4*M.
INODE, JNODE	Each is an integer vector of length M, INODE(i), JNODE(i) are the end nodes of the ith edge in the connected undirected graph.

Output:

KCONCT	The edge-connectivity of the graph.

Working storages:

For the description of the following working arrays, see the parameters of subroutine NFLOW.

IEDGE	Integer vector of length M4.
JEDGE	Integer vector of length M4.
CAPAC	Integer vector of length M4.
MINCUT	Integer vector of length N.
FLOW	Integer vector of length M4.
NODFLO	Integer vector of length N.
POINT	Integer vector of length N.
IMAP	Integer vector of length N.
JMAP	Integer vector of length N.

Subroutine NFLOW parameters

Let G be a network of *E* edges.

Input:

N	Number of nodes.

M — Equal to 2^*E.

INODE, JNODE — Each is an integer vector of length M; an edge in G directed from node u to node v will be represented by two directed edges (u, v) and (v, u), where

$$\text{INODE}(i) = \text{u}, \quad \text{JNODE}(i) = \text{v},$$
$$\text{INODE}(j) = \text{u}, \quad \text{JNODE}(j) = \text{v},$$

for some i and j. On output, the edges will be sorted lexicographically.

CAPAC — Integer vector of length M; CAPAC(i) is the edge capacity of edge (u, v) in G, and the artificially created edge (v, u) will have an edge capacity CAPAC(j) equal to zero.

ISORCE, ISINK — A maximum flow is required from node ISORCE to node ISINK in the network.

Output:

MINCUT — Integer vector of length N; MINCUT(i) = 1 if node i is in the minimal cut set; otherwise, it is equal to zero.

FLOW — Integer vector of length M; FLOW(i) is the amount of flow on edge i.

NODFLO — Integer vector of length N; NODFLO(i) is the amount of flow through node i.

Working storages:

POINT — Integer vector of length N; POINT(i) is the first edge from node i.

IMAP — Integer vector of length N; pointer array.

JMAP — Integer vector of length N; pointer array.

REMARK. As an example for using NFLOW, we want to find the maximum flow from node 3 to node 2 in the following network of $E = 5$ edges.

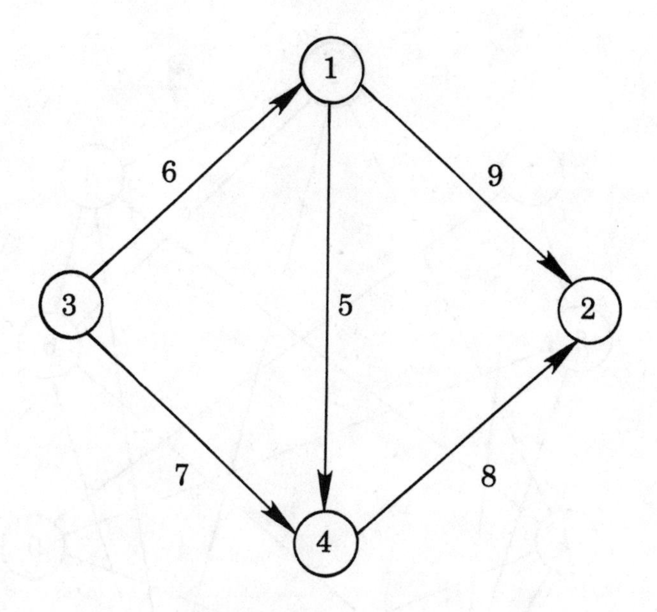

The numbers on the edges represent the edge capacity.

The input data to subroutine NFLOW might be:

N = 4
M = 10
INODE: 4 2 3 1 1 2 3 4 1 4
JNODE: 2 4 1 3 2 1 4 3 4 1
CAPAC: 8 0 6 0 9 0 7 0 5 0
ISORCE = 3
ISINK = 2

Notice that the edges can be arranged in an arbitrary order.

D. Test example

Find the edge-connectivity of the following graph with nine nodes and 17 edges.

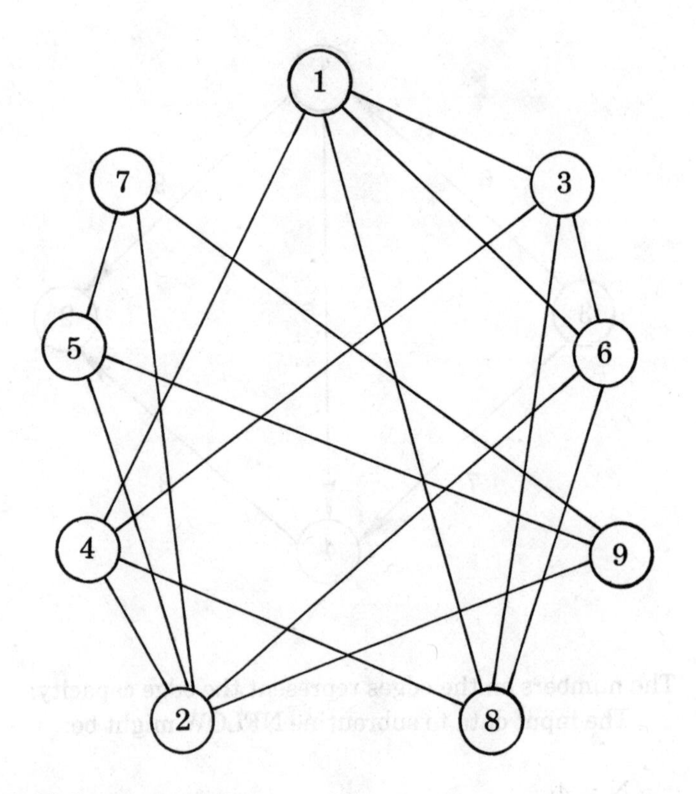

E. Program listing

MAIN PROGRAM

```
      INTEGER INODE(17),JNODE(17),IEDGE(68),
     +           JEDGE(68),CAPAC(68),MINCUT(9),
     +           FLOW(68),NODFLO(9),POINT(9),
     +           IMAP(9),JMAP(9)
      DATA INODE / 6,2,3,6,7,1,4,7,3,4,9,6,5,4,2,9,4/,
     +     JNODE / 8,5,1,3,2,8,3,5,8,1,2,1,9,8,6,7,2/
C
      N = 9
      M = 17
      M4 = 4*M
      CALL EDGECN(N,M,M4,INODE,JNODE,
     +            KCONCT,IEDGE,JEDGE,CAPAC,
     +            MINCUT,FLOW,NODFLO,
     +            POINT,IMAP,JMAP)
```

```
      WRITE(*,10) KCONCT
10    FORMAT(' THE EDGE CONNECTIVITY OF',
     +   ' THE INPUT GRAPH IS',I4)
      STOP
      END
```

OUTPUT RESULT

```
 THE EDGE CONNECTIVITY OF THE INPUT GRAPH
   IS 2
```

```
      SUBROUTINE EDGECN(N,M,M4,INODE,
     +                          JNODE,KCONCT,
     +                          IEDGE,JEDGE,
     +                          CAPAC,MINCUT,
     +                          FLOW, NODFLO,
     +                          POINT,IMAP,JMAP)
C
C     Find the edge-connectivity of a connected graph
C
      INTEGER INODE(M),JNODE(M),IEDGE(M4),
     +             JEDGE(M4),CAPAC(M4),
     +             MINCUT(N),FLOW(M4),
     +             NODFLO(N),POINT(N),IMAP(N),
     +             JMAP(N)
C
C     Duplicate the edges
C
      J = 0
      DO 10 I = 1, M
        J = J + 1
        IEDGE(J) = INODE(I)
        JEDGE(J) = JNODE(I)
        CAPAC(J) = 1
        J = J + 1
        IEDGE(J) = JNODE(I)
        JEDGE(J) = INODE(I)
        CAPAC(J) = 0
        J = J + 1
        IEDGE(J) = JNODE(I)
```

```
        JEDGE(J) = INODE(I)
        CAPAC(J) = 1
        J = J + 1
        IEDGE(J) = INODE(I)
        JEDGE(J) = JNODE(I)
        CAPAC(J) = 0
 10   CONTINUE
C
C     Call on the network flow algorithm
C
      KCONCT = N
      ISORCE = 1
      DO 20 ISINK = 2, N
        CALL NFLOW(N,M4,IEDGE,JEDGE,
     +             CAPAC,ISORCE,ISINK,
     +             MINCUT,FLOW,NODFLO,
     +             POINT,IMAP,JMAP)
        IF (NODFLO(ISORCE) .LT. KCONCT)
     +    KCONCT = NODFLO(ISORCE)
 20   CONTINUE
C
      RETURN
      END
      SUBROUTINE NFLOW(N,M,INODE,
     +                 JNODE,CAPAC,
     +                 ISORCE,ISINK,
     +                 MINCUT,FLOW,
     +                 NODFLO,POINT,
     +                 IMAP,JMAP)
C
C     Karzanov's network flow algorithm
C
      INTEGER INODE (M),JNODE(M),
     +        CAPAC(M),MINCUT(N),
     +        FLOW(M),NODFLO(N),
     +        POINT(N),IMAP(N),JMAP(N)
      LOGICAL FINISH
C
```

```
C         Initialize
C
          DO 10 I = 1, N
   10       POINT(I) = 0
          IFLOW = 0
          DO 20 I = 1, M
            FLOW(I) = 0
            J = INODE(I)
            IF (J .EQ. ISORCE)
     +        IFLOW = IFLOW + CAPAC(I)
            POINT(J) = POINT(J) + 1
   20     CONTINUE
          NODFLO(ISORCE) = IFLOW
          NODEW = 1
          DO 30 I = 1, N
            J = POINT(I)
            POINT(I) = NODEW
            IMAP(I) = NODEW
            NODEW = NODEW + J
   30     CONTINUE
          FINISH = .FALSE.
C
C         Sort the edges in lexicographical order
C
   40     IFLAG = 0
   50     IF (IFLAG .LT. 0) THEN
            IF (IFLAG .NE. - 1) THEN
              IF (NODEW .LT. 0) NODEI = NODEI + 1
              NODEJ = JCONT
              JCONT = NODEI
              IFLAG = -1
            ELSE
              IF (NODEW .LE. 0) THEN
                IF (ICONT .GT. 1) THEN
                  ICONT = ICONT - 1
                  JCONT = ICONT
                  GOTO 60
                ENDIF
```

```
            IF (M1 .EQ. 1) THEN
              IFLAG = 0
            ELSE
              NODEI = M1
              M1 = M1 - 1
              NODEJ = 1
              IFLAG = 1
            ENDIF
          ELSE
            IFLAG = 2
          ENDIF
        ENDIF
      ELSE
        IF (IFLAG .GT. 0) THEN
          IF (IFLAG .LE. 1) JCONT = ICONT
          GOTO 60
        ELSE
          M1 = M
          ICONT = 1 + M / 2
          ICONT = ICONT - 1
          JCONT = ICONT
60        NODEI = JCONT + JCONT
          IF (NODEI .LT. M1) THEN
            NODEJ = NODEI + 1
            IFLAG = -2
          ELSE
            IF (NODEI .EQ. M1) THEN
              NODEJ = JCONT
              JCONT = NODEI
              IFLAG = -1
            ELSE
              IF (ICONT .GT. 1) THEN
                ICONT = ICONT - 1
                JCONT = ICONT
                GOTO 60
              ENDIF
              IF (M1 .EQ. 1) THEN
```

```
                  IFLAG = 0
                ELSE
                  NODEI = M1
                  M1 = M1 - 1
                  NODEJ = 1
                  IFLAG = 1
                ENDIF
              ENDIF
            ENDIF
          ENDIF
        ENDIF
        IF (IFLAG .LT. 0) THEN
          NODEW = INODE(NODEI)
     +       - INODE(NODEJ)
          IF (NODEW .EQ. 0)
            NODEW = JNODE(NODEI)
     +       - JNODE(NODEJ)
          GOTO 50
        ELSE
          IF (IFLAG .GT. 0) THEN
C
C           Interchange two edges
C
  70        NODEW = INODE(NODEI)
            INODE(NODEI) = INODE(NODEJ)
            INODE(NODEJ) = NODEW
            IFLOW = CAPAC(NODEI)
            CAPAC(NODEI) = CAPAC(NODEJ)
            CAPAC(NODEJ) = IFLOW
            NODEW = JNODE(NODEI)
            JNODE(NODEI) = JNODE(NODEJ)
            JNODE(NODEJ) = NODEW
            IFLOW = FLOW(NODEI)
            FLOW(NODEI) = FLOW(NODEJ)
            FLOW(NODEJ) = IFLOW
            IF (IFLAG .GT. 0) GOTO 50
            IF (IFLAG .EQ. 0) GOTO 160
```

```
            JMAP(NODEV) = NODEJ
            GOTO 240
         ELSE
            IF (FINISH) RETURN
         ENDIF
      ENDIF
C
C     Set the cross references between edges
C
      DO 80 I = 1, M
         NODEV = JNODE(I)
         INODE(I) = IMAP(NODEV)
         IMAP(NODEV) = IMAP(NODEV) + 1
   80 CONTINUE
C
   90 IFLAG = 0
      DO 100 I = 1, N
         IF (I .NE. ISORCE) NODFLO(I) = 0
         JMAP(I) = M + 1
         IF (I .LT. N) JMAP(I) = POINT (I + 1)
         MINCUT(I) = 0
  100 CONTINUE
C
      IN = 0
      IOUT = 1
      IMAP(1) = ISORCE
      MINCUT(ISORCE) = - 1
  110 IN = IN + 1
      IF (IN .LE. IOUT) THEN
         NODEU = IMAP(IN)
         MEDGE = JMAP(NODEU) - 1
         IEND = POINT(NODEU) - 1
  120    IEND = IEND + 1
            IF (IEND .GT. MEDGE) GOTO 110
            NODEV = JNODE(IEND)
            IFLOW = CAPAC(IEND) - FLOW(IEND)
         IF ((MINCUT(NODEV) .NE. 0) .OR.
     +      (IFLOW .EQ. 0)) GOTO 120
         IF (NODEV .NE. ISINK) THEN
```

```
          IOUT = IOUT + 1
          IMAP(IOUT) = NODEV
        ENDIF
        MINCUT(NODEV) = -1
        GOTO 120
      ENDIF
      IF (MINCUT(ISINK) .EQ. 0) THEN
C
C       Exit
C
        DO 130 I = 1, N
 130      MINCUT(I) = -MINCUT(I)
        DO 140 I = 1, M
          NODEU = JNODE(INODE(I))
          IF (FLOW(I) .LT. 0)
     +              NODFLO(NODEU)
     +              = NODFLO(NODEU)
     +              - FLOW(I)
          INODE(I) = NODEU
 140    CONTINUE
        NODFLO(ISORCE) = NODFLO(ISINK)
        FINISH = .TRUE.
        GOTO 40
      ENDIF
      MINCUT(ISINK) = 1
 150  IN = IN - 1
      IF (IN .NE. 0) THEN
        NODEU = IMAP(IN)
        NODEI = POINT(NODEU) - 1
        NODEJ = JMAP(NODEU) - 1
 160    IF (NODEI .NE. NODEJ) THEN
          NODEV = JNODE(NODEJ)
          IF ((MINCUT(NODEV) .LE. 0) .OR.
     +         (CAPAC(NODEJ) .EQ.
     +          FLOW(NODEJ))) THEN
            NODEJ = NODEJ - 1
            GOTO 160
          ENDIF
          JNODE(NODEJ) = -NODEV
```

```
          CAPAC(NODEJ) = CAPAC(NODEJ)
     +      - FLOW(NODEJ)
          FLOW(NODEJ) = 0
          NODEI = NODEI + 1
          IF (NODEI .LT. NODEJ) THEN
            INODE(INODE(NODEI)) = NODEJ
            INODE(INODE(NODEJ)) = NODEI
            GOTO 70
          ENDIF
        ENDIF
        IF (NODEI .GE. POINT(NODEU))
     +    MINCUT(NODEU) = NODEI
        GOTO 150
      ENDIF
      NODEX = 0
      DO 170 I = 1, IOUT
        IF (MINCUT(IMAP(I)) .GT. 0) THEN
          NODEX = NODEX + 1
          IMAP(NODEX) = IMAP(I)
        ENDIF
170   CONTINUE
C
C     Find a feasible flow
C
      IFLAG = -1
      NODEY = 1
180   NODEU = IMAP(NODEY)
      IF (NODFLO(NODEU) .LE. 0) THEN
190     NODEY = NODEY + 1
        IF (NODEY .LE. NODEX) GOTO 180
        IPARM = 0
200     NODEY = NODEY - 1
        IF (NODEY .NE. 1) THEN
          NODEU = IMAP(NODEY)
          IF (NODFLO(NODEU) .LT. 0) GOTO 200
          IF (NODFLO(NODEU) .EQ. 0) THEN
C
C             Accumulating flows
C
```

```
          MEDGE = M + 1
          IF (NODEU .LT. N)
     +       MEDGE = POINT(NODEU + 1)
          IEND = JMAP(NODEU)
          JMAP(NODEU) = MEDGE
210       IF (IEND .EQ. MEDGE) GOTO 200
             J = INODE(IEND)
             IFLOW = FLOW(J)
             FLOW(J) = 0
             CAPAC(J) = CAPAC(J) - IFLOW
             FLOW(IEND) = FLOW(IEND)
     +          - IFLOW
             IEND = IEND + 1
          GOTO 210
        ENDIF
        IF (POINT(NODEU) .GT.
     +    MINCUT(NODEU)) THEN
          IEND = JMAP(NODEU)
          GOTO 250
        ENDIF
        IEND = MINCUT(NODEU) + 1
        GOTO 230
      ENDIF
      DO 220 I = 1, M
        NODEV = -JNODE(I)
        IF (NODEV .LT. 0) GOTO 220
          JNODE(I) = NODEV
          J = INODE(I)
          CAPAC(I) = CAPAC(I) - FLOW(J)
          IFLOW = FLOW(I) - FLOW(J)
          FLOW(I) = IFLOW
          FLOW(J) = -IFLOW
220   CONTINUE
      GOTO90
    ENDIF
C
C   An outgoing edge from a node is given
C     maximum flow
C
```

```
      IEND = MINCUT(NODEU) + 1
230   IEND = IEND - 1
      IF (IEND .GE. POINT(NODEU)) THEN
        NODEV = -JNODE(IEND)
        IF (NODFLO(NODEV) .LT. 0) GOTO 230
        IFLOW = CAPAC(IEND) - FLOW(IEND)
        IF (NODFLO(NODEU) .LT. IFLOW)
     +    IFLOW = NODFLO(NODEU)
        FLOW(IEND) = FLOW(IEND) + IFLOW
        NODFLO(NODEU) = NODFLO(NODEU)
     +    - IFLOW
        NODFLO(NODEV) = NODFLO(NODEV)
     +    + IFLOW
        IPARM = 1
        NODEI = INODE(IEND)
        NODEJ = JMAP(NODEV) - 1
        IF (NODEI .LT. NODEJ) THEN
          INODE(INODE(NODEI)) = NODEJ
          INODE(INODE(NODEJ)) = NODEI
          GOTO 70
        ENDIF
        IF (NODEI .EQ. NODEJ)
     +    JMAP(NODEV) = NODEJ
240     IF (NODFLO(NODEU) .GT. 0) GOTO 230
        IF (CAPAC(IEND) .EQ. FLOW(IEND))
     +    IEND = IEND - 1
      ENDIF
      MINCUT(NODEU) = IEND
      IF (IPARM .NE. 0) GOTO 190
C
C     Remove excess incoming flows from nodes
C
      IEND = JMAP(NODEU)
250   J = INODE(IEND)
        IFLOW = FLOW(J)
        IF (NODFLO(NODEU) .LT. IFLOW)
     +    IFLOW = NODFLO(NODEU)
        FLOW(J) = FLOW(J) - IFLOW
```

```
            NODFLO(NODEU) = NODFLO(NODEU)
        +     - IFLOW
            NODEV = JNODE(IEND)
            NODFLO(NODEV) - NODFLO(NODEV)
        +     + IFLOW
            IEND = IEND + 1
          IF (NODFLO(NODEU) .GT. 0) GOTO 250
          NODFLO(NODEU) = -1
          GOTO 200
C
          END
```

1-3 Fundamental Set of Cycles

A. Problem description

Let T be a spanning tree of an undirected graph G. The *fundamental set of cycles* of G corresponding to T is the set of cycles of G consisting of one edge (i, j) of G-T together with the unique path between node i and node j in T. The problem is to find a fundamental set of cycles in a given undirected graph that is not necessarily connected.

B. Method

Let G be the given undirected graph of n nodes. First, find all the connected components of G. Then, the fundamental set of cycles of G can be found for each component H of G as follows.

STEP 1. Let E be the set of edges and V the set of nodes of H. Take any node v from V as the root of the tree consisting of the single node. Set

$T = \{v\}, S = V.$

STEP 2. Let x be any node in $T \cap S$. If such a node does not exist, then stop.

STEP 3. Consider each edge (x, y) in E.

If y is in T, then generate the fundamental cycle consisting of edge (x, y) together with the unique path between x and y in the tree, and delete the edge (x, y) from E.

If y is not in T, then add the edge (x, y) to the tree, add the node y to T, and delete the edge (x, y) from E.

STEP 4. Remove the node x from S and return to Step 2.

The processing time of the algorithm is $O(n^2)$.

C. Subroutine FCYCLE parameters

Input:

N	Number of nodes.
M	Number of edges.
INODE, JNODE	Each is an integer vector of length M; INODE(i), JNODE(i) are the end nodes of the ith edge in the input graph.

Output:

NUMCYC	Number of independent cycles in the graph.
NUMCMP	Number of components of the graph.
NCYC	Integer vector of length N; each time the output WRITE statement in subroutine FCYCLE is executed, a fundamental cycle NCYC(i), i = 1, 2, . . . , LEN is generated, where LEN is a local variable in FCYCLE.

Working storages:

FWDARC	Integer vector of length M; FWDARC(i) is the ending node of the ith edge in the forward star representation of the graph.
ARCFIR	Integer vector of length N; ARCFIR(i) is

	the number of the first edge starting at node i in the forward star representation.
NEXT	Integer vector of length N; NEXT(i) is the number of the node following node i on the path from node i to the root of the tree.
IPOINT	Integer vector of length N; pointer array.

D. Test example

Find a fundamental set of cycles for the following graph.

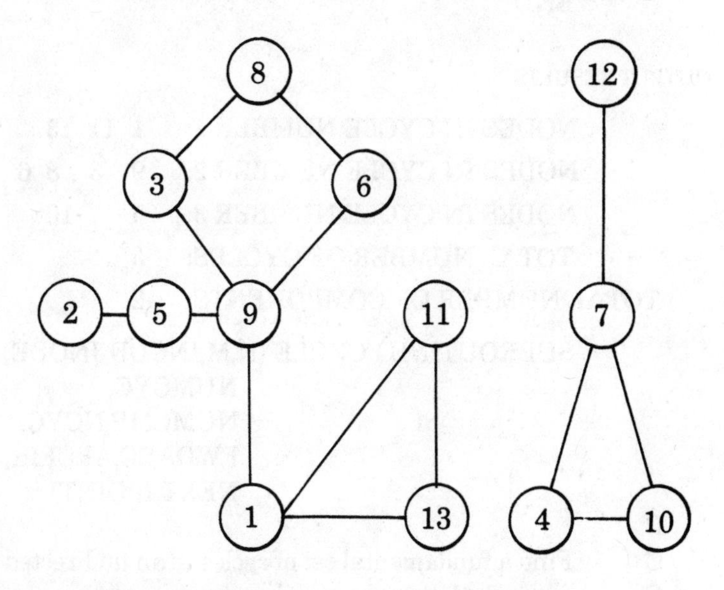

E. Program listing

MAIN PROGRAM

```
      INTEGER INODE(14),JNODE(14),
     +          FWDARC(14),ARCFIR(13),
     +          NCYC(13),NEXT(13),IPOINT(13)
      DATA INODE /5,4,11,6,5, 7, 1,8,9,10,3, 1, 4,6/,
     +     JNODE /9,7,13,8,2,12,11,3,1, 7,9,13,10,9/
C
      N = 13
      M = 14
```

```
      CALL FCYCLE (N,M,INODE,JNODE,
     +             NUMCYC,NUMCMP,NCYC,
     +             FWDARC,ARCFIR,NEXT,
     +             IPOINT)
      WRITE(*,10) NUMCYC,NUMCMP
   10 FORMAT(//5X,'TOTAL NUMBER OF',
     +           ' CYCLES:',I3//
     +           ' TOTAL NUMBER OF',
     +           ' COMPONENTS:',I3)
      STOP
      END
```

OUTPUT RESULTS

```
      NODES IN CYCLE NUMBER 1:    1  11  13
      NODES IN CYCLE NUMBER 2:    9   3   8  6
      NODES IN CYCLE NUMBER 3:    4   7  10
      TOTAL NUMBER OF CYCLES:     3
  TOTAL NUMBER OF COMPONENTS:     2
```

```
      SUBROUTINE FCYCLE (N,M,INODE,JNODE,
     +                   NUMCYC,
     +                   NUMCMP,NCYC,
     +                   FWDARC,ARCFIR,
     +                   NEXT,IPOINT)
C
C     Find a fundamental set of cycles in an undirected
C       graph that is not necessarily connected
C
      INTEGER INODE(M),JNODE(M),NCYC(N),
     +        FWDARC(M),ARCFIR(N),NEXT(N),
     +        IPOINT(N)
      LOGICAL JOIN
C
C     Set up the forward star representation of the
C       graph
C
      N1 = N - 1
      K = 0
```

```
      DO 20 I = 1, N1
        ARCFIR(I) = K + 1
        DO 10 J = 1, M
          NODEU = INODE(J)
          NODEV = JNODE(J)
          IF ((NODEU .EQ. I) .AND.
     +        (NODEU .LT. NODEV)) THEN
            K = K + 1
            FWDARC(K) = NODEV
          ELSE
            IF ((NODEV .EQ. I) .AND.
     +          (NODEV .LT. NODEU)) THEN
              K = K + 1
              FWDARC(K) = NODEU
            END IF
          END IF
   10   CONTINUE
   20 CONTINUE
      ARCFIR(N) = M + 1
C
      DO 30 IROOT = 1, N
   30   NEXT(IROOT) = 0
      NUMCMP = 0
      NUMCYC = 0
C
      DO 100 IROOT = 1, N
        IF (NEXT(IROOT) .EQ. 0) THEN
          NUMCMP = NUMCMP + 1
          NEXT(IROOT) = -1
          INDEX = 1
          IPOINT(1) = IROOT
          IEDGE = 2
C
   40     NODE3 = IPOINT(INDEX)
          INDEX = INDEX - 1
          NEXT(NODE3) = -NEXT(NODE3)
C
          DO 90 NODE2 = 1, N
            JOIN = .FALSE.
```

```
              IF (NODE2 .NE. NODE3) THEN
                IF (NODE2 .LT. NODE3) THEN
                  NODEU = NODE2
                  NODEV = NODE3
                ELSE
                  NODEU = NODE3
                  NODEV = NODE2
                ENDIF
                LOW = ARCFIR(NODEU)
                IUP = ARCFIR(NODEU + 1)
                IF (IUP .GT. LOW) THEN
                  IUP = IUP - 1
                  DO 50 K = LOW, IUP
                    IF (FWDARC(K) .EQ. NODEV)
     +                THEN
                      JOIN = .TRUE.
                      GOTO 60
                    ENDIF
   50             CONTINUE
                ENDIF
              ENDIF
   60         IF (JOIN) THEN
                NODE1 = NEXT(NODE2)
                IF (NODE1 .EQ. 0) THEN
                  NEXT(NODE2) = -NODE3
                  INDEX = INDEX + 1
                  IPOINT(INDEX) = NODE2
                ELSE
                  IF (NODE1 .LT. 0) THEN
C
C                   Generate the next cycle
C
                    NUMCYC = NUMCYC + 1
                    LEN = 3
                    NODE1 = -NODE1
                    NCYC(1) = NODE1
                    NCYC(2) = NODE2
                    NCYC(3) = NODE3
                    I = NODE3
```

```
 70                J = NEXT(I)
                   IF (J .NE. NODE1) THEN
                     LEN = LEN + 1
                     NCYC(LEN) = J
                     I = J
                     GOTO 70
                   ENDIF
C
C                  Output a cycle
C
                   WRITE(*,80) NUMCYC,(NCYC(I),
     +               I=1,LEN)
 80                FORMAT(/' NODES IN CYCLE',
     +                    ' NUMBER',I3),
     +                    ':',1X,20I4)
                 ENDIF
               ENDIF
             ENDIF
 90        CONTINUE
C
           IEDGE = IEDGE + 1
           IF ((IEDGE .LE. N)
     +       .AND. (INDEX .GT. 0)) GOTO 40
C
           NEXT(IROOT) = 0
           NODE3 = IPOINT(1)
           NEXT(NODE3) = IABS(NEXT(NODE3))
         ENDIF
100    CONTINUE
C
       RETURN
       END
```

1-4 Cut Nodes and Bridges

A. Problem description

Find all the cut nodes and bridges of a given undirected graph that is not necessarily connected.

B. Method

Assume that input graph G is connected and has the set of nodes V numbered from 1 to n. A tree T rooted at an arbitrary node r will be grown to span G. The unique predecessor of each node i in the tree is denoted by $p(i)$. Let $d(j)$ be the distance from node i to the root of T, $b(i)$ be the label assigned to the edge $(i, p(i))$, and $h(i)$ be a Boolean variable for marking edge i.

STEP 1. Set

$$b(i) = 0, h(i) = \text{FALSE}$$

for all i. Choose an arbitrary node r as the root.

Initially, T consists of the single node r,

$$d(r) = 0, X \text{ is empty}, Y = T, Z = V - \{r\}.$$

STEP 2. If Y is nonempty then select the most recent member of Y, say u, delete u from Y and continue from Step 3.

If Y is empty then count the number of blocks of G by noting that edge $(i, p(i))$ belongs to block $b(i)$; moreover, if $b(i)$ is equal to $-i$, then the edge itself is a block of G. If G has only one block then stop; otherwise, G has at least one cut node. The cut nodes are identified by the property that node i is a cut node if and only if there are two or more distinct labels on edges of T through i. Stop.

STEP 3. Set $L = 0$.
For each edge (u, v), where v is in Z, do the following:

add (u, v) to T and transfer v from Z to Y,
set $p(v) = u, d(v) = d(u) + 1, b(v) = -v$.

For each edge (u, v), where v is in Y, do the following:

if $b(v) > 0$ then set $h(b(v)) = \text{TRUE}$,
set $b(v) = u$ and $L = \max(L, d(u) - d(p(v)))$.

STEP 4. If $L = 0$ then add u to X and return to Step 2. Otherwise, for each edge, say $(i, p(i))$, on the path of length L from u to the root of the tree, set

$$h(b(i)) = \text{TRUE and } b(i) = u.$$

For each edge $(j, p(j))$ for which $h(b(j)) = \text{TRUE}$, set

$$b(j) = u.$$

Add u to X and return to Step 2.

The running time of the algorithm is bound by $O(n^2)$.

C. Subroutine CONECT parameters

Input:

N	Number of nodes.
M	Number of edges.
INODE, JNODE	Each is an integer vector of length M; INODE(i), JNODE(i) are the end nodes of the ith edge in the input graph.
IROOT	A specified node for the root of the tree.

Output:

NCUT	Number of cut nodes.
NBRIDG	Number of bridges.
CUTNOD	Integer vector of length N; during execution of the subroutine, CUTNOD(i) is the unique predecessor of node i in the tree. On output, the cut nodes are stored in CUTNODE(i), $i = 1, 2, \ldots$, NCUT.
BRIDGE	Integer vector of length M; during execution of the subroutine, BRIDGE(i) is

the end node of the ith edge in the forward star representation of the graph. On output, if BRIDGE (i) = 1 then the edge (INODE(i),JNODE(i)) is a bridge, and if BRIDGE(i) = 0 then the ith edge is not a bridge.

Working storages:

ARCFIR	Integer vector of length N; ARCFIR(i) is the number of the first edge starting at node i in the forward star representation of the graph.
LABEL	Integer vector of length N; labels assigned to edges in the tree.
NEXT	Integer vector of length N; NEXT(i) is the number of the node following node i on the path from node i to the root of the tree. On input, NEXT(i) must be initialized to zero, $i = 1, 2, \ldots, n$.
LENGTH	Integer vector of length N; LENGTH(i) is the distance from node i to the root of the tree.
NEW	Logical vector of length N; indicates whether the labels are replaced.

REMARK. If the input graph is known to be connected, then one single call of subroutine CONECT rooted at an arbitrary node will find all the cut nodes and bridges of the graph.

If it is not known whether the input graph is connected, then the calling of subroutine CONECT can be enclosed in a loop iterating from IROOT = 1 to N. As usual, the array NEXT should be initialized to zero at the start of the loop. Each execution of CONECT will assign a positive integer to NEXT(i), for every node i in the same component of IROOT. If there are k components in the input graph, the loop will be executed k times. This is illustrated in the test example below.

D. Test example

Find all the cut nodes and bridges of the following graph.

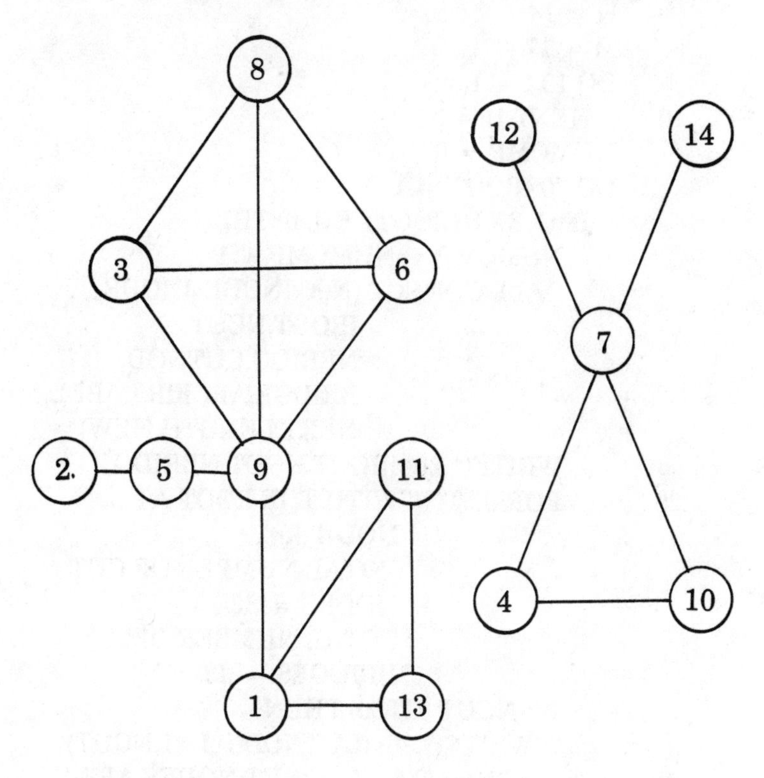

E. Program listing

MAIN PROGRAM

```
      INTEGER INODE(17),JNODE(17),
     +        BRIDGE(17),CUTNOD(14),
     +        ARCFIR(14),LABEL(14),
     +        NEXT(14),LENGTH(14)
      LOGICAL NEW(14)
      DATA INODE /3, 4,5,13, 7,9,14, 1,3,9, 7,
     +  9,5,9, 1,6,4/,
```

```
     +      JNODE /6,10,9,11,12,8, 7,11,8,1,10,
     +   3,2,6,13,8,7/
C
      N = 14
      M = 17
      DO 10 I = 1, N
 10     NEXT(I) = 0
      NUMCMP = 0
      DO 70 IROOT = 1, N
        IF (NEXT(IROOT) .EQ. 0) THEN
          NUMCMP = NUMCMP + 1
          CALL CONECT (N,M,INODE,JNODE,
     +                 IROOT,NCUT,
     +                 NBRIDG,CUTNOD,
     +                 BRIDGE,ARCFIR,LABEL,
     +                 NEXT,LENGTH,NEW)
          WRITE(*,20) IROOT,NCUT,NBRIDG
 20       FORMAT(///' WITH THE ROOT AT',
     +           ' NODE',I3,','/
     +           ' TOTAL NUMBER OF CUT',
     +           ' NODES =',I3/
     +           ' TOTAL NUMBER OF',
     +           ' BRIDGES =',I3)
          IF (NCUT .GT. 0) THEN
            WRITE(*,30) (CUTNOD(I),I=1,NCUT)
 30         FORMAT(/'    THE CUT NODES ARE:',
     +        10I5)
          ENDIF
          IF (NBRIDG .GT. 0) THEN
            WRITE(*,40)
 40         FORMAT(/'       THE BRIDGES ARE:')
            DO 60 I = 1, M
              IF (BRIDGE(I) .GT. 0) THEN
                WRITE(*,50) INODE(I),JNODE(I)
 50             FORMAT(6X,2I4)
              ENDIF
 60         CONTINUE
          ENDIF
        ENDIF
```

```
 70     CONTINUE
        WRITE(*,80) NUMCMP
 80     FORMAT(//' TOTAL NUMBER OF',
      +   ' COMPONENTS = ',I3)
        STOP
        END
```

OUTPUT RESULTS

```
WITH THE ROOT AT NODE 1,
  TOTAL NUMBER OF CUT NODES = 3
    TOTAL NUMBER OF BRIDGES = 3
  THE CUT NODES ARE: 1   5   9
  THE BRIDGES ARE:
        5  9
        9  1
        5  2
WITH THE ROOT AT NODE 4,
  TOTAL NUMBER OF CUT NODES = 1
    TOTAL NUMBER OF BRIDGES = 2
  THE CUT NODES ARE: 7
  THE BRIDGES ARE:
         7  12
        14   7
TOTAL NUMBER OF COMPONENTS = 2
```

```
        SUBROUTINE CONECT (N,M,INODE,
      +                    JNODE,IROOT,
      +                    NCUT,NBRIDG,
      +                    CUTNOD,BRIDGE,
      +                    ARCFIR,LABEL,
      +                    NEXT,LENGTH,
      +                    NEW)
C
C       Find the bridges, blocks and cut nodes of an
C         undirected graph
C
```

```
      INTEGER INODE(M),JNODE(M),
     +          CUTNOD(N),BRIDGE(M),
     +          ARCFIR(N),LABEL(N),NEXT(N),
     +          LENGTH(N)
      LOGICAL NEW(N),JOIN
C
C     Set up the forward star representation of the
C       graph
C
      N1 = N - 1
      K = 0
      DO 20 I = 1, N1
        ARCFIR(I) = K + 1
        DO 10 J = 1, M
          NODEU = INODE(J)
          NODEV = JNODE(J)
          IF ((NODEU .EQ. I) .AND.
     +      (NODEU .LT. NODEV)) THEN
            K = K + 1
            BRIDGE(K) = NODEV
          ELSE
            IF ((NODEV .EQ. I) .AND.
     +        (NODEV .LT. NODEU)) THEN
              K = K + 1
              BRIDGE(K) = NODEU
            ENDIF
          ENDIF
 10     CONTINUE
 20   CONTINUE
      ARCFIR(N) = M + 1
C
      DO 30 I = 1, N
        LABEL(I) = 0
 30     NEW(I) = .FALSE.
C
      LENGTH(IROOT) = 0
      NEXT(IROOT) = -1
```

```
      LABEL(IROOT) = -IROOT
      INDEX = 1
      CUTNOD(1) = IROOT
      IEDGE = 2
C
   40 NODE3 = CUTNOD(INDEX)
      INDEX = INDEX - 1
      NEXT(NODE3) = -NEXT(NODE3)
      LEN1 = 0
C
      DO 70 NODE2 = 1, N
        JOIN = .FALSE.
        IF (NODE2 .NE. NODE3) THEN
          IF (NODE2 .LT. NODE3) THEN
            NODEU = NODE2
            NODEV = NODE3
          ELSE
            NODEU = NODE3
            NODEV = NODE2
          ENDIF
          LOW = ARCFIR(NODEU)
          IUP = ARCFIR(NODEU + 1)
          IF (IUP .GT. LOW) THEN
            IUP = IUP - 1
            DO 50 K = LOW, IUP
              IF (BRIDGE(K) .EQ. NODEV) THEN
                JOIN = .TRUE.
                GOTO 60
              ENDIF
   50       CONTINUE
          ENDIF
        ENDIF
   60   IF (JOIN) THEN
          NODE1 = NEXT(NODE2)
          IF (NODE1 .EQ. 0) THEN
            NEXT(NODE2) = -NODE3
            INDEX = INDEX + 1
```

```
              CUTNOD(INDEX) = NODE2
              LENGTH(NODE2) = LENGTH(NODE3)
     +          + 1
              LABEL(NODE2) = -NODE2
            ELSE
              IF (NODE1 .LT. 0) THEN
C
C               Next block
C
                NODE4 = LABEL(NODE2)
                IF (NODE4 .GT. 0) NEW(NODE4)
     +             = .TRUE.
                LABEL(NODE2) = NODE3
                LEN2 = LENGTH(NODE3)
     +             - LENGTH(-NODE1)
                IF (LEN2 .GT. LEN1) LEN1 = LEN2
              ENDIF
            ENDIF
          ENDIF
   70   CONTINUE
C
        IF (LEN1 .GT. 0) THEN
          J = NODE3
   80     LEN1 = LEN1 - 1
          IF (LEN1 .GE. 0) THEN
            ITEMP = LABEL(J)
            IF (ITEMP .GT. 0) NEW(ITEMP) = .TRUE.
            LABEL(J) = NODE3
            J = NEXT(J)
            GOTO 80
          ENDIF
          DO 90 I = 1, N
            ITEMP = LABEL(I)
            IF (ITEMP .GT. 0) THEN
              IF (NEW(ITEMP)) LABEL(I) = NODE3
            ENDIF
   90     CONTINUE
        ENDIF
C
```

```
      IEDGE = IEDGE + 1
      IF ((IEDGE .LE. N) .AND. (INDEX .GT. 0))
     +   GOTO 40
C
      NEXT(IROOT) = 0
      NODE3 = CUTNOD(1)
      NEXT(NODE3) = IABS(NEXT(NODE3))
      NBRIDG = 0
      NBLOCK = 0
      NCUT = 0
      DO 110 I = 1, N
        IF (I .NE. IROOT) THEN
          NODE3 = LABEL(I)
          IF (NODE3 .LT. 0) THEN
            NBLOCK = NBLOCK + 1
            NBRIDG = NBRIDG + 1
            LABEL(I) = N + NBLOCK
          ELSE
            IF ((NODE3 .LE. N) .AND.
     +          (NODE3 .GT. 0)) THEN
              NBLOCK = NBLOCK + 1
              NODE4 = N + NBLOCK
              DO 100 J = I, N
                IF (LABEL(J) .EQ. NODE3)
     +            LABEL(J) = NODE4
  100         CONTINUE
            ENDIF
          ENDIF
        ENDIF
  110 CONTINUE
      DO 120 I = 1, N
        ITEMP = LABEL(I)
        IF (ITEMP .GT. 0) LABEL(I) = ITEMP - N
  120 CONTINUE
C
      I = 1
  130 IF (NEXT(I) .NE. IROOT) THEN
        I = I + 1
        GOTO 130
```

```
      ENDIF
      LABEL(IROOT) = LABEL(I)
      DO 140 I = 1, N
        NODE1 = NEXT(I)
        IF (NODE1 .GT. 0) THEN
          ITEMP = IABS(LABEL(NODE1))
          IF (IABS(LABEL(I)) .NE. ITEMP)
     +      LABEL(NODE1) = -ITEMP
        ENDIF
  140 CONTINUE
      DO 150 I = 1, N
        IF (LABEL(I) .LT. 0) NCUT = NCUT + 1
  150 CONTINUE
C
C     Store the cut nodes
C
      J = 0
      DO 160 I = 1, N
        IF (LABEL(I) .LT. 0) THEN
          J = J + 1
          CUTNODE(J) = I
        ENDIF
  160 CONTINUE
C
C     Find the end nodes
C
      DO 170 I = 1, N
  170   LENGTH(I) = 0
      DO 180 I = 1, M
        J = INODE(I)
        LENGTH(J) = LENGTH(J) + 1
        J = JNODE(I)
        LENGTH(J) = LENGTH(J) + 1
  180 CONTINUE
      DO 190 I = 1, N
        IF (LENGTH(I) .EQ. 1) THEN
          IF (LABEL(I) .GT. 0)
     +      LABEL(I) = -LABEL(I)
        ENDIF
```

```
190   CONTINUE
C
C     Store the bridges
C
      DO 200 I = 1, M
200     BRIDGE(I) = 0
      DO 230 I = 1, N
        IF (LABEL(I) .LT. 0) THEN
          DO 220 J = 1, M
            NODEV = 0
            IF (I .EQ. INODE(J))
     +        NODEV = JNODE(J)
            IF (I .EQ. JNODE(J))
     +        NODEV = INODE(J)
            IF (NODEV .GT. 0) THEN
              IF (LABEL(NODEV) .LT. 0) THEN
                IF (LABEL(I) .NE. LABEL(NODEV))
     +            THEN
                  BRIDGE(J) = 1
                ELSE
                  K = -LABEL(I)
                  DO 210 II = 1, N
                    IF ((II .NE. I) .AND.
     +                  (II .NE. NODEV)) THEN
                      IF (LABEL(II) .EQ. K)
     +                  GOTO 230
                    ENDIF
210               CONTINUE
                  BRIDGE(J) = 1
                ENDIF
              ENDIF
            ENDIF
220       CONTINUE
        ENDIF
230   CONTINUE
C
      RETURN
      END
```

1-5 Strongly Connected Components

A. Problem description

A strongly connected component of a directed graph is a maximal set of nodes in which there is a directed path from any one node in the set to any other node in the set. The problem is to find the strongly connected components of a given directed graph.

B. Method

A depth-first search will be used to find the strongly connected components of a directed graph G of n nodes and m edges.

STEP 1. A depth-first search of G is performed by selecting one node s of G as a start node; s is marked "visited." Each unvisited node adjacent to s is searched in turn, using the depth-first search recursively. The search of s is completed when all nodes that can be reached from s have been visited. If some nodes remain unvisited, then an unvisited node is arbitrarily selected as a new start node. This process is repeated until all nodes of G have been visited.

STEP 2. Let $r_1, r_2, \ldots, r_k$ be the roots in the order in which the depth-first search of the nodes terminated. Then, the strongly connected component G_1 with root r_1 consists of all descendants of r_1. Furthermore, for each i, $i = 2, 3, \ldots, k$, the strongly connected component G_i with root r_i consists of these nodes which are descendants of r_i but are in none of $G_1, G_2, \ldots, G_{i-1}$.

The running time of the algorithm is $O(\max(n, m))$.

C. Subroutine STCOMP parameters

Input:

N	Number of nodes.
M	Number of edges.
N1	Equal to N + 1.
INODE, JNODE	Each is an integer vector of length M; the *i*th edge of the input graph is directed from node INODE(*i*) to node JNODE(*i*). The graph does not necessarily have to be connected.

Output:

NUMCOM	Number of strongly connected components of the graph.
NODESC	Integer vector of length N; containing the node numbers belonging to the successive strongly connected components of the graph, the node number of the first node of each component is negative.

Working storages:

FWDARC	Integer vector of length M; FWDARC(*i*) is the ending node of the *i*th edge in the forward star representation of the graph.
ARCFIR	Integer vector of length N1; ARCFIR(*i*) + 1 is the number of the first edge starting at node *i* in the forward star representation of the graph ARCFIR(1) = 0.
NUMBER	Integer vector of length N; array for the order of the nodes as they are visited.
LOWLNK	Integer vector of length N; LOWLNK(*i*) is the smallest node which is in the same component as node *i*.

STACK1	Integer vector of length N; stack of node numbers.
STACK2	Integer vector of length N; stack of node numbers.

D. Test example

Find the strongly connected components of the following graph.

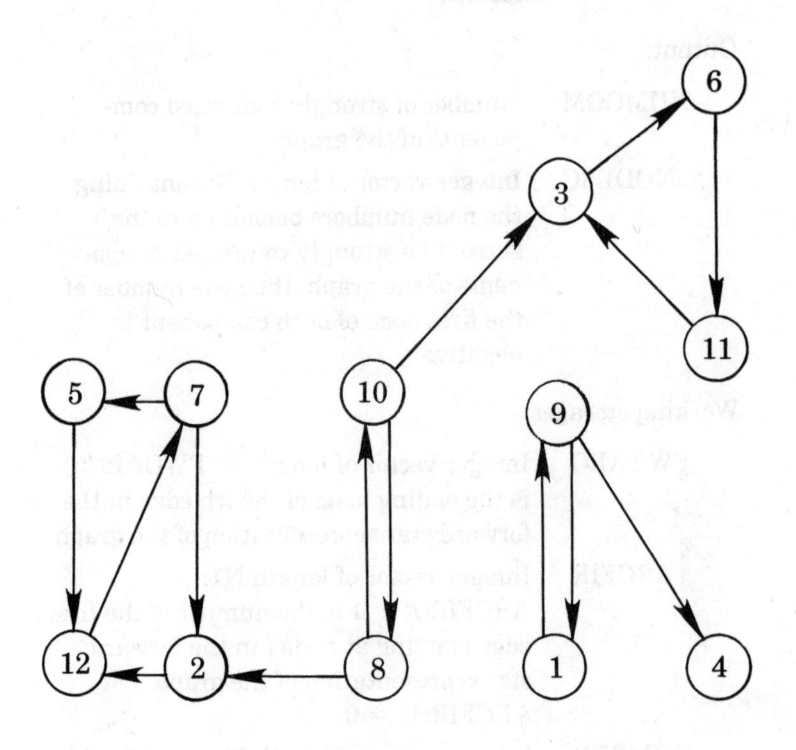

E. Program listing

MAIN PROGRAM

```
      INTEGER INODE(15),JNODE(15),
     +        NODESC(12),FWDARC(15),
     +        ARCFIR(13),NUMBER(12),
```

```
     +               LOWLNK(12),STACK1(12),
     +               STACK2(12)
      DATA INODE /8,11,9, 5,3, 8,9, 6,12,10,
     +          7,1, 2,10,7/,
     +        JNODE /2, 3,1,12,6,10,4,11, 7, 3,
     +          2,9,12, 8,5/
C
      N = 12
      M = 15
      N1 = N + 1
      CALL STCOMP (N,M,N1,INODE,JNODE,
     +                     NUMCOM,NODESC,
     +                     FWDARC,ARCFIR,
     +                     NUMBER,LOWLNK,
     +                     STACK1,STACK2)
      WRITE(*,10) NUMCOM
   10 FORMAT(' THERE ARE',I2,' STRONGLY',
     +              ' CONNECTED COMPONENTS.'///
     +            '   THE NODES IN COMPONENT',
     +                ' 1 are:'/)
      K = 1
      DO 40 I = 1, N
        J = NODESC(I)
        IF (J .LT. 0) THEN
          WRITE(*,20) -J
   20     FORMAT(10X,I6)
          IF (K .LT. NUMCOM) THEN
            K = K + 1
            WRITE(*,30) K
   30       FORMAT(/' THE NODES IN',
     +          ' COMPONENT',I2,' ARE:'/)
          ENDIF
        ELSE
          WRITE(*,20) J
        ENDIF
   40 CONTINUE
C
      STOP
      END
```

OUTPUT RESULTS

```
THERE ARE 5 STRONGLY CONNECTED
   COMPONENTS.
THE NODES IN COMPONENT 1 ARE:
                    4
THE NODES IN COMPONENT 2 ARE:
                    9
                    1
THE NODES IN COMPONENT 3 ARE:
                    5
                    7
                   12
                    2
THE NODES IN COMPONENT 4 ARE:
                   11
                    6
                    3
THE NODES IN COMPONENT 5 ARE:
                   10
                    8
```

```
      SUBROUTINE STCOMP (N,M,N1,INODE,
     +                   JNODE,NUMCOM,
     +                   NODESC,FWDARC,
     +                   ARCFIR,NUMBER,
     +                   LOWLNK,STACK1,
     +                   STACK2)
C
C     Find the strongly connected components of a
C        directed graph
C
      INTEGER INODE(M),JNODE(M),NODESC(N),
     +        FWDARC(M),ARCFIR(N1),
     +        NUMBER(N),LOWLNK(N),
     +        STACK1(N),STACK2(N)
C
```

```
C       Set up the forward star representation of the
C         graph
C
        ARCFIR(1) = 0
        K = 0
        DO 20 I = 1, N
          DO 10 J = 1, M
            IF (INODE(J) .EQ. I) THEN
              K = K + 1
              FWDARC(K) = JNODE(J)
            ENDIF
  10      CONTINUE
          ARCFIR(I + 1) = K
  20    CONTINUE
C
        DO 30 I = 1, N
  30      NUMBER(I) = 0
        J1 = 0
        NUMCOM = 0
        J3 = 0
        J2 = 0
C
        DO 80 NODEI = 1, N
          IF (NUMBER(NODEI) .EQ. 0) THEN
            ISTACK = 1
            STACK2(1) = NODEI
  40        NODEV = STACK2(ISTACK)
            J1 = J1 + 1
            NUMBER(NODEV) = J1
            LOWLNK(NODEV) = J1
            J2 = J2 + 1
            STACK1(J2) = NODEV
            INDEX3 = ARCFIR(NODEV + 1)
            INDEX = ARCFIR(NODEV) + 1
  50        IF (INDEX .LE. INDEX3) THEN
              NODEU = FWDARC(INDEX)
              INDEX2 = NUMBER(NODEU)
              IF (INDEX2 .EQ. 0) THEN
                ISTACK = ISTACK + 1
```

```
          STACK2(ISTACK) = NODEU
          GOTO 40
        ELSE
          IF (INDEX2 .LT. NUMBER(NODEV))
     +      THEN
            DO 60 I = 1, J2
              IF (STACK1(I) .EQ. NODEU)
     +          THEN
                IF (INDEX2 .LT.
     +            LOWLNK(NODEV))
     +            LOWLNK(NODEV)
     +              = INDEX2
                INDEX = INDEX + 1
                GOTO 50
              ENDIF
   60       CONTINUE
          ENDIF
        ENDIF
        INDEX = INDEX + 1
        GOTO 50
      ENDIF
C
      INDEX1 = NUMBER(NODEV)
      IF (LOWLNK(NODEV) .EQ. INDEX1)
     +  THEN
        NUMCOM = NUMCOM + 1
        J4 = J2
   70   IF (J4 .GT. 0) THEN
          INDEX2 = STACK1(J4)
          IF (NUMBER(INDEX2) .GE. INDEX1)
     +      THEN
            J3 = J3 + 1
            NODESC(J3) = INDEX2
            J4 = J4 - 1
            GOTO 70
          ENDIF
        ENDIF
        NODESC(J3) = -NODESC(J3)
        J2 = J4
```

```
            ENDIF
            IF (ISTACK .GT. 1) THEN
               NODEU = STACK2(ISTACK)
               ISTACK = ISTACK - 1
               NODEV = STACK2(ISTACK)
               INDEX1 = LOWLNK(NODEU)
               IF (INDEX1 .LT. LOWLNK(NODEV))
     +            LOWLNK(NODEV) = INDEX1
               INDEX = INDEX + 1
               GOTO 50
            ENDIF
         ENDIF
   80 CONTINUE
C
      RETURN
      END
```

1-6 Minimal Equivalent Graph

A. Problem description

The minimal equivalent graph problem is to find a directed graph H from a given strongly connected graph G by removing the maximum number of edges from G without affecting its reachability properties. That is, for any two nodes u and v in G, there is a directed path from node u to node v in H.

B. Method

Let E be the set of edges in a given strongly connected graph G of n nodes and m edges.

STEP 1. Examine all edges sequentially. An edge (i, j) is removed from G whenever there exists an alternative path from node i to node j which does not include previously eliminated edges.

Let F be the set of edges which are removed from G, and $S = E - F$ be the current solution.

STEP 2. Iteratively execute a combination of backtracking and forward moves. A backtracking move will remove from F the highest labelled edge—say the kth one—and add it back in E. This backtracking move is followed by a forward move which will sequentially consider all edges in E with labels greater than k, removing from E every edge (i, j) for which an alternative path exists, and adding (i, j) back to F.

STEP 3. IF $|E| < |S|$, then update the current solution by setting $S = E$, and return to Step 2.

The algorithm terminates when no further backtracking is possible in Step 2 (in which case F is empty) or when $|S| = n$.

The number of iterations of the algorithm is bounded by $O(2^m)$.

C. Subroutine MINEQV parameters

Input:

N	Number of nodes.
M	Number of edges.
N1	Equal to N + 1.
INODE, JNODE	Each is an integer vector of length M; the ith edge of the input graph is directed from node INODE(i) to node JNODE(i). It is assumed that the input graph is strongly connected.

Output:

ARCLIS	Logical vector of length M; ARCLIS(*i*) has the value TRUE if the *i*th edge is in the minimal equivalent graph; otherwise, it has the value FALSE.

Working storages:

FWDARC	Integer vector of length M; FWDARC(*i*) is the ending node of the *i*th edge in the forward star representation of the graph.
ARCFIR	Integer vector of length N1; ARCFIR(*i*) is the number of the first edge starting at node *i* in the forward star representation of the graph.
POINT	Integer vector of length M; pointing to the original edge list in the forward star representation.
MARK	Logical vector of length M; a working copy of the array ARCLIS.
IDESCN	Integer vector of length N; IDESCN(*i*) is the number of descendants of node *i*.
IANCES	Integer vector of length N; IANCES(*i*) is the number of ancestors of node *i*.
NOPATH	Logical vector of length N; NOPATH(*i*) indicates whether there is a path from the current node to node *i*.
NEXT	Integer vector of length N; the next nodes to be investigated.

D. Test example

Find a minimal equivalent graph of the following strongly connected graph.

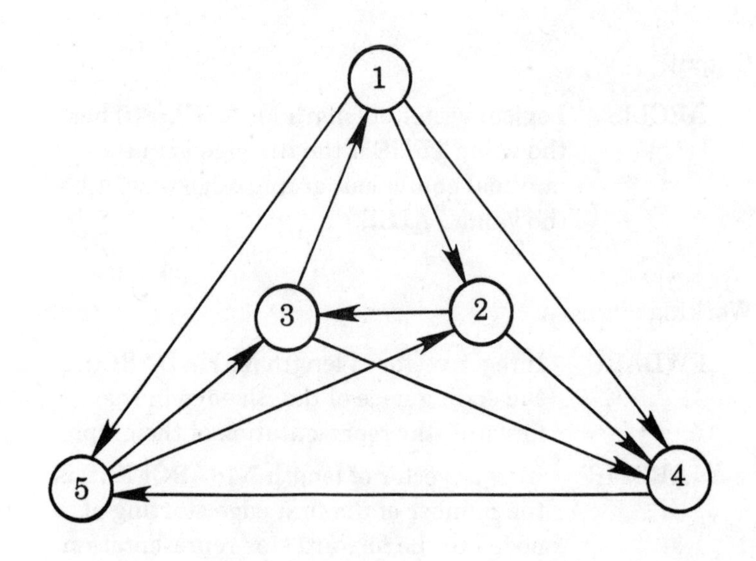

E. Program listing

MAIN PROGRAM

```
      INTEGER INODE(10),JNODE(10),
     +          FWDARC(10),ARCFIR(6),
     +          POINT(10),IDESCN(5),
     +          IANCES(5),NEXT(5)
      LOGICAL ARCLIS(10),MARK(10),NOPATH(5)
      DATA INODE /5,2,2,4,1,3,1,5,1,3/,
     +      JNODE /2,3,4,5,5,4,2,3,4,1/
      N = 5
      M = 10
      N1 = N + 1
      CALL MINEQV (N,M,N1,INODE,JNODE,
     +              ARCLIS,FWDARC,ARCFIR,
     +              POINT,MARK,IDESCN,
     +              IANCES,NOPATH,NEXT)
      WRITE(*,10)
   10 FORMAT(' THE EDGES OF THE MINIMAL',
     +   ' EQUIVALENT GRAPH:'/)
      DO 30 I = 1, M
         IF (ARCLIS(I)) THEN
            WRITE(*,20) INODE(I),JNODE(I)
```

```
20          FORMAT(14X,2I3)
          ENDIF
30     CONTINUE
       STOP
       END
```

OUTPUT RESULTS

```
THE EDGES OF THE MINIMAL EQUIVALENT
   GRAPH:
                        5 2
                        2 3
                        4 5
                        1 5
                        3 4
                        3 1
```

```
       SUBROUTINE MINEQV (N,M,N1,INODE,
     +                   JNODE,ARCLIS,
     +                   FWDARC,ARCFIR,
     +                   POINT,MARK,
     +                   IDESCN,IANCES,
     +                   NOPATH,NEXT)
C
C      Find a minimal equivalent graph of a strongly
C        connected digraph
C
       INTEGER INODE(M),JNODE(M),
     +         FWDARC(M),ARCFIR(N1),
     +         POINT(M),IDESCN(N),
     +         IANCES(N),NEXT(N)
       LOGICAL ARCLIS(M),MARK(M),NOPATH(N),
     +   PEXIST,JOIN
C
C      Set up the forward star representation of the
C        graph
C
       K = 0
```

```
      DO 20 I = 1, N
        ARCFIR(I) = K + 1
        DO 10 J = 1, M
          IF (INODE(J) .EQ. I) THEN
            K = K + 1
            POINT(K) = J
            FWDARC(K) = JNODE(J)
            MARK(K) = .TRUE.
          ENDIF
10      CONTINUE
20    CONTINUE
      ARCFIR(N1) = M + 1
C
C     Compute the number of descendants and
C        ancestors of each node
C
      DO 30 I = 1, N
        IDESCN(I) = 0
30      IANCES(I) = 0
      IEDGES = 0
      DO 40 K = 1, M
        I = INODE(K)
        J = JNODE(K)
        IDESCN(I) = IDESCN(I) + 1
        IANCES(J) = IANCES(J) + 1
        IEDGES = IEDGES + 1
40    CONTINUE
      IF (IEDGES .EQ. N) THEN
        DO 50 K = 1, M
50        ARCLIS(POINT(K)) = MARK(K)
        RETURN
      ENDIF
      IERASE = 0
      DO 60 K = 1, M
        I = INODE(POINT(K))
        J = JNODE(POINT(K))
C
C     Check for the existence of an alternative path
C
```

```
      IF (IDESCN(I) .NE. 1) THEN
        IF (IANCES(J) .NE. 1) THEN
          MARK(K) = .FALSE.
          CALL FINDP (N,M,N1,I,J,FWDARC,
     +                    ARCFIR,MARK,
     +                    PEXIST,NEXT,NOPATH)
            IF (PEXIST) THEN
              IDESCN(I) = IDESCN(I) - 1
              IANCES(J) = IANCES(J) - 1
              IERASE = IERASE + 1
            ELSE
              MARK(K) = .TRUE.
            ENDIF
          ENDIF
        ENDIF
   60 CONTINUE
C
      IF (IERASE .EQ. 0) THEN
        DO 70 K = 1, M
   70     ARCLIS(POINT(K)) = MARK(K)
        RETURN
      ENDIF
C
      IHIGH = 0
      I1 = N
      J1 = N
C
C     Store the current best solution
C
   80 DO 90 K = 1, M
   90   ARCLIS(K) = MARK(K)
      JERASE = IERASE
      IF (IEDGES - JERASE .EQ. N) THEN
        DO 100 K = 1, M
  100     MARK(K) = ARCLIS(K)
        DO 110 K = 1, M
  110     ARCLIS(POINT(K)) = MARK(K)
        RETURN
      ENDIF
```

```
C
C       Forward move
C
 120    JOIN = .FALSE.
        LOW = ARCFIR(I1)
        IUP = ARCFIR(I1 + 1)
        IF (IUP .GT. LOW) THEN
          IUP = IUP - 1
          DO 130 K = LOW, IUP
            IF (FWDARC(K) .EQ. J1) THEN
              JOIN = .TRUE.
              KEDGE = K
              GOTO 140
            ENDIF
 130      CONTINUE
        ENDIF
C
 140    IF (JOIN) THEN
          IF (.NOT. MARK(KEDGE)) THEN
            MARK(KEDGE) = .TRUE.
            IDESCN(I1) = IDESCN(I1) + 1
            IANCES(J1) = IANCES(J1) + 1
            IERASE = IERASE - 1
            IF (IERASE + IHIGH - (N - I1)
     +          .GT. JERASE) GOTO 220
          ENDIF
          IHIGH = IHIGH + 1
        ENDIF
C
 150    IF (J1 .NE. 1) THEN
          J1 = J1 - 1
          GOTO 120
        ENDIF
C
        IF (I1 .EQ. 1) THEN
          DO 160 K = 1, M
 160        MARK(K) = ARCLIS(K)
          DO 170 K = 1, M
```

```
170         ARCLIS(POINT(K)) = MARK(K)
          RETURN
        ENDIF
        I1 = I1 - 1
        J1 = N
        GOTO 120
C
C       Backtrack move
C
180     JOIN = .FALSE.
        LOW = ARCFIR(I1)
        IUP = ARCFIR(I1 + 1)
        IF (IUP .GT. LOW) THEN
          IUP = IUP - 1
          DO 190 K = LOW, IUP
            IF (FWDARC(K) .EQ. J1) THEN
              JOIN = .TRUE.
              KEDGE = K
              GOTO 200
            ENDIF
190       CONTINUE
        ENDIF
C
200     IF (JOIN) THEN
          IHIGH= I HIGH - 1
          IF (IDESCN(I1) .NE. 1) THEN
            IF (IANCES(J1) .NE. 1) THEN
              MARK(KEDGE) = .FALSE.
              CALL FINDP (N,M,N1,I1,J1,FWDARC,
     +                    ARCFIR,MARK,
     +                    PEXIST,NEXT,NOPATH)
              IF (PEXIST) THEN
                IDESCN(I1) = IDESCN(I1) - 1
                IANCES(J1) = IANCES(J1) - 1
                IERASE = IERASE + 1
                GOTO 210
              ENDIF
              MARK(KEDGE) = .TRUE.
```

```
        ENDIF
      ENDIF
C
      IF (IERASE + IHIGH - (N - I1)
   +     .LE. JERASE) THEN
        IHIGH = IHIGH + 1
        GOTO 150
      ENDIF
C
C     Check for the termination of the forward move
C
210   IF (IHIGH - (N - I1) .EQ. 0) GOTO 80
      ENDIF
C
220   IF (J1 .NE. N) THEN
        J1 = J1 + 1
        GOTO 180
      ENDIF
      IF (I1 .NE. N) THEN
        I1 = I1 + 1
        J1 = 1
        GOTO 180
      ENDIF
      GOTO 80
C
      END
      SUBROUTINE FINDP (N,M,N1,I1,J1,
   +                    FWDARC,ARCFIR,
   +                    MARK,PEXIST,
   +                    NEXT,NOPATH)
C
C     Determine if a path exists from I1 to J1 by
C       Yen's algorithm (this subprogram is used by
C       subroutine MINEQV)
C
      INTEGER FWDARC(M),ARCFIR(N1),
   +   NEXT(N)
```

```
      LOGICAL MARK(M),NOPATH(N),PEXIST,
     +  JOIN
C
C     Initialize
C
      DO 10 I = 1, N
        NEXT(I) = I
 10     NOPATH(I) = .FALSE.
      NOPATH(I1) = .TRUE.
      NEXT(I1) = N
      INDEX1 = I1
      INDEX2 = N - 1
C
C     Compute the shortest distance labels
C
      I = 1
 20     J = NEXT(I)
        JOIN = .FALSE.
        LOW = ARCFIR(INDEX1)
        IUP = ARCFIR(INDEX1 + 1)
        IF (IUP .GT. LOW) THEN
          IUP = IUP - 1
          DO 30 K = LOW, IUP
            IF (FWDARC(K) .EQ. J) THEN
              JOIN = .TRUE.
              KEDGE = K
              GOTO 40
            ENDIF
 30       CONTINUE
        ENDIF
C
 40     IF (JOIN) THEN
          IF (MARK(KEDGE))
     +      NOPATH(J) = .TRUE.
        ENDIF
        IF (NOPATH(J)) GOTO 50
        I = I + 1
```

```
         IF (I .LE. INDEX2) GOTO 20
        PEXIST = .FALSE.
        RETURN
C
 50     INDEX3 = I + 1
        IF (INDEX3 .LE. INDEX2) THEN
        DO 80 I2 = INDEX3, INDEX2
          J2 = NEXT(I2)
          JOIN = .FALSE.
          LOW = ARCFIR(INDEX1)
          IUP = ARCFIR(INDEX1 + 1)
          IF (IUP .GT. LOW) THEN
            IUP = IUP - 1
            DO 60 K = LOW, IUP
              IF (FWDARC(K) .EQ. J2) THEN
                JOIN = .TRUE.
                KEDGE = K
                GOTO 70
              ENDIF
 60         CONTINUE
          ENDIF
 70       IF (JOIN) THEN
            IF (MARK(KEDGE))
     +        NOPATH(J2) = .TRUE.
          ENDIF
 80     CONTINUE
        ENDIF
C
C       Check whether an alternative path exists
C
        IF (NOPATH(J1)) THEN
          PEXIST = .TRUE.
          RETURN
        ENDIF
        NEXT(I) = NEXT(INDEX2)
        INDEX1 = J
        INDEX2 = INDEX2 - 1
        IF (INDEX2 .GT. 1) GOTO 20
          JOIN = .FALSE.
```

```
             LOW = ARCFIR(INDEX1)
             IUP = ARCFIR(INDEX1 + 1)
             IF (IUP .GT. LOW) THEN
               IUP = IUP - 1
               DO 90 K = LOW, IUP
                 IF (FWDARC(K) .EQ. J1) THEN
                   JOIN = .TRUE.
                   KEDGE = K
                   GOTO 100
                 ENDIF
   90          CONTINUE
             ENDIF
  100      PEXIST = .FALSE.
           IF (JOIN) THEN
             IF (MARK(KEDGE)) PEXIST = .TRUE.
           ENDIF
C
           RETURN
           END
```

1-7 Maximal Independent Sets

A. Problem description

Consider an undirected graph G. An *independent set* is a subset of nodes of G such that no two nodes of the set are adjacent in G. An independent set is maximal if there is no other independent set that contains it. A *clique* is a subset of nodes of G in which every two nodes in the set are adjacent in G.

The problem is to find all the maximal independent sets and cliques of a given undirected graph.

B. Method

It is clear that a subset of nodes C of a graph G is a maximal independent set if and only if C is a clique in the complement of G. Thus, any algorithm which finds the maximal independent sets of a graph can also be used to find its cliques, and vice versa.

The algorithm for finding all maximal independent sets is essentially an enumerative tree search. Let $P(j)$ be an independent set at stage j, and $Q(j)$ be the largest set of nodes such that any node from $Q(j)$ added to $P(j)$ will produce an independent set $P(j+1)$. The set $Q(j)$ can be partitioned into two disjoint sets $S(j)$ and $T(j)$, where $S(j)$ is the set of all nodes which have been used in the search to augment $P(j)$, and $T(j)$ is the set of all nodes which have not been used.

Let $E(u)$ be the set of nodes v such that node u and node v are adjacent in G. The algorithm can be described as follows.

STEP 1. Set $P(0)$ = empty, $S(0)$ = empty, $T(0)$ = set of nodes of G, and $j = 0$.

STEP 2. Perform a forward branch in the tree search by choosing any node $x(j)$ from $T(j)$, and create three new sets:

$$P(j+1) = P(j) + \{x(j)\},$$
$$S(j+1) = S(j) - E(x(j)),$$
$$T(j+1) = T(j) - E(x(j)) - \{x(j)\}.$$

Set $j = j + 1$.

STEP 3. If there exists a node y in $S(j)$ such that $E(y)$ is disjoint from $T(j)$, then go to Step 5.

STEP 4. If both $S(j)$ and $T(j)$ are empty, then output the maximal independent set $P(j)$ and go to Step 5.

If $S(j)$ is nonempty and $T(j)$ is empty, then go to Step 5; otherwise, return to Step 2.

STEP 5. Backtrack by setting $j = j - 1$.

Remove node $x(j)$ from $P(j+1)$ to produce $P(j)$. Remove node $x(j)$ from $T(j)$ and add it to $S(j)$.

IF $j = 0$ and $T(0)$ is empty, then stop; otherwise, return to Step 3.

C. Subroutine CLIQUE parameters

Input:

N	Number of nodes.
M	Number of edges.
N1	Equal to N + 1.
M2	Equal to M + M.
INODE, JNODE	Each is an integer vector of length M; INODE(i) and JNODE(i) are the two end nodes of the ith edge in the input graph which is not necessarily connected.

Output:

CLIQ — Integer vector of length N; each time the output WRITE statement in subroutine CLIQUE is executed, a clique

$$\text{CLIQ}(i),\ i = 1, 2, \ldots, \text{NUM},$$

is generated, where NUM is a local variable in CLIQUE.

Working storages:

FWDARC	Integer vector of length M2; FWDARC(i) is the ending node of the ith edge in the forward star representation of the graph.
ARCFIR	Integer vector of length N1; ARCFIR(i) is the number of the first edge starting at node i in the forward star representation of the graph.
CAND1	Integer vector of length N1; positions of candidates which have not been used.
CAND2	Integer vector of length N1; positions of candidates which have been used.
STACK	Integer matrix of dimension N1 by N1; stack of nodes at different stages.

NDPS1	Integer vector of length N; array of nodes at each stage.
NDPS2	Integer vector of length N; array of nodes at each stage.

D. Test example

Generate all cliques of the following graph.

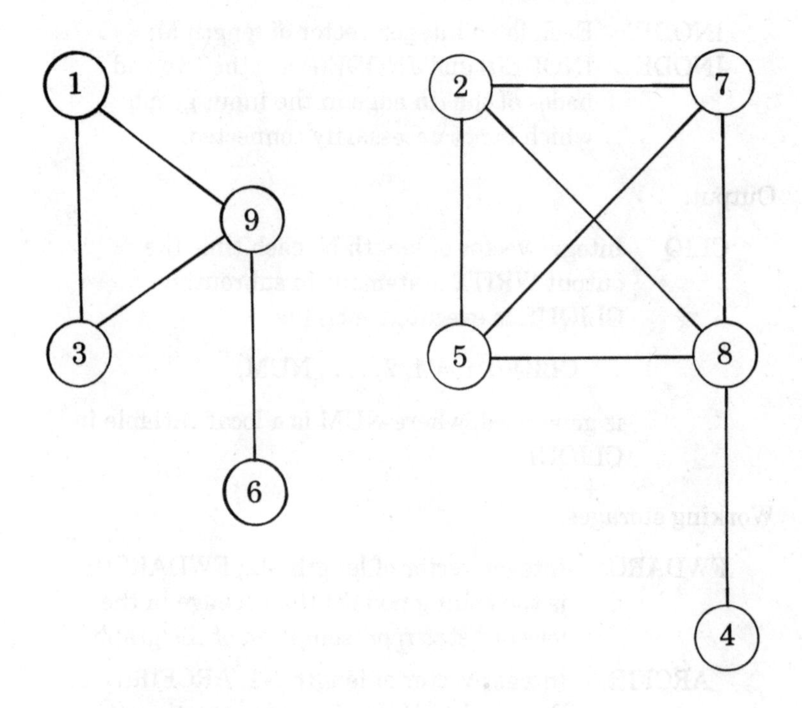

E. Program listing

MAIN PROGRAM

```
      INTEGER INODE(11),JNODE(11),CLIQ(9)
     +        FWDARC(22),ARCFIR(10),CAND1(10),
     +        CAND2(10),NDPS1(9),NDPS2(9),
     +        STACK(10,10)
      DATA INODE /5,1,3,6,2,5,5,1,2,7,8/,
```

```
+         JNODE /8,9,9,9,7,2,7,3,8,8,4/
  N = 9
  M = 11
  N1 = N + 1
  M2 = M + M
  CALL CLIQUE (N,M,N1,M2,INODE,JNODE,CLIQ,
+                  FWDARC,ARCFIR,CAND1,
+                  CAND2,STACK,NDPS1,NDPS2)
  STOP
  END
```

OUTPUT RESULTS

```
CLIQUE:    8 2 5 7
CLIQUE:    8 4
CLIQUE:    3 1 9
CLIQUE:    6 9
```

```
      SUBROUTINE CLIQUE (N,M,N1,M2,INODE,
     +                   JNODE,CLIQ,
     +                   FWDARC,ARCFIR,
     +                   CAND1,CAND2,
     +                   STACK,NDPS1,
     +                   NDPS2)
C
C     Find all the cliques of an undirected graph
C
      INTEGER INODE(M),JNODE(M),CLIQ(N),
     +          FWDARC(M2),ARCFIR(N1),
     +          CAND1(N1),CAND2(N1),
     +          STACK(N1,N1),NDPS1(N),
     +          NDPS2(N)
      LOGICAL JOIN
C
C     Set up the foward star representation of the
C       graph
C
      K = 0
      DO 20 I = 1, N
        ARCFIR(I) = K + 1
```

```
      DO 10 J = 1, M
        IF(INODE(J) .EQ. I) THEN
          K = K + 1
          FWDARC(K) = JNODE(J)
        ENDIF
        FI (JNODE(J) .EQ. I) THEN
          K = K + 1
          FWDARC(K) = INODE(J)
        ENDIF
 10     CONTINUE
 20   CONTINUE
      ARCFIR(N1) = K + 1
C
      LEVEL = 1
      LEVEL1 = 2
      DO 30 I = 1, N
 30   STACK(LEVEL,I) = I
      NUM = 0
      CAND2(LEVEL) = 0
      CAND1(LEVEL) = N
C
 40   ISMALL = CAND1(LEVEL)
      NODEU = 0
      NDPS1(LEVEL) = 0
C
 50   NODEU = NODEU + 1
      IF ((NODEU .LE. CAND1(LEVEL)) .AND.
     +  (ISMALL .NE. 0)) THEN
        ISUB1 = STACK(LEVEL,NODEU)
        ISUM = 0
        NODEV = CAND2(LEVEL)
C
 60   NODEV = NODEV + 1
      IF ((NODEV .LE. CAND1(LEVEL)) .AND.
     +  (ISUM .LT. ISMALL)) THEN
        ITEMP = STACK(LEVEL,NODEV)
        IF (ITEMP .EQ. ISUB1)THEN
          JOIN = .TRUE.
        ELSE
```

```
          JOIN = .FALSE.
          LOW = ARCFIR(ITEMP)
          IUP = ARCFIR(ITEMP + 1)
            IF (IUP .GT. LOW) THEN
              IUP = IUP - 1
              DO 70 K = LOW, IUP
                IF (FWDARC(K) .EQ. ISUB1) THEN
                  JOIN = .TRUE.
                  GOTO 80
                ENDIF
 70           CONTINUE
            ENDIF
          ENDIF
C
C         Store the potential candidate
C
 80       IF (.NOT. JOIN) THEN
            ISUM = ISUM + 1
            INDEXV = NODEV
          ENDIF
          GOTO 60
        ENDIF
 C
        IF (ISUM .LT. ISMALL) THEN
          NDPS2(LEVEL) = ISUB1
          ISMALL = ISUM
          IF (NODEU .LE. CAND2(LEVEL)) THEN
            NODEW = INDEXV
          ELSE
            NODEW = NODEU
            NDPS1(LEVEL) = 1
          ENDIF
        ENDIF
        GOTO 50
      ENDIF
C
C     Backtrack
C
      NDPS1(LEVEL) = ISMALL + NDPS1(LEVEL)
```

```
 90   IF (NDPS1(LEVEL) .GT. 0) THEN
        ISUB1 = STACK(LEVEL,NODEW)
        STACK(LEVEL,NODEW)
     +     = STACK(LEVEL,CAND2(LEVEL)+1)
        STACK(LEVEL,CAND2(LEVEL)+1)
     +     = ISUB1
        ISUB2 = ISUB1
        NODEU = 0
        CAND2(LEVEL1) = 0
C
100     NODEU = NODEU + 1
        IF (NODEU .LE. CAND2(LEVEL)) THEN
          ITEMP = STACK(LEVEL,NODEU)
          IF (ITEMP .EQ. ISUB2) THEN
            JOIN = .TRUE.
          ELSE
            JOIN = .FALSE.
            LOW = ARCFIR(ITEMP)
            IUP = ARCFIR(ITEMP + 1)
            IF (IUP .GT. LOW) THEN
              IUP = IUP - 1
              DO 110 K = LOW, IUP
                IF (FWDARC(K) .EQ. ISUB2) THEN
                  JOIN = .TRUE.
                  GOTO 120
                ENDIF
110           CONTINUE
            ENDIF
          ENDIF
C
120       IF (JOIN) THEN
            CAND2(LEVEL1)
     +        = CAND2(LEVEL1) + 1
            STACK(LEVEL1,CAND2(LEVEL1))
     +        = STACK(LEVEL,NODEU)
          ENDIF
          GOTO 100
        ENDIF
C
```

```
      CAND1(LEVEL1) = CAND2(LEVEL1)
      NODEU = CAND2(LEVEL) + 1
130   NODEU = NODEU + 1
      IF (NODEU .LE. CAND1(LEVEL)) THEN
        ITEMP = STACK(LEVEL,NODEU)
C
        IF (ITEMP .EQ. ISUB2) THEN
          JOIN = .TRUE.
        ELSE
          JOIN = .FALSE.
          LOW = ARCFIR(ITEMP)
          IUP = ARCFIR(ITEMP + 1)
          IF (IUP .GT. LOW) THEN
            IUP = IUP - 1
            DO 140 K = LOW, IUP
              IF (FWDARC(K) .EQ. ISUB2) THEN
                JOIN = .TRUE.
                GOTO 150
              ENDIF
140         CONTINUE
          ENDIF
        ENDIF
C
150     IF (JOIN) THEN
          CAND1(LEVEL1)
     +      = CAND1(LEVEL1) + 1
          STACK(LEVEL1,CAND1(LEVEL1))
     +      = STACK(LEVEL,NODEU)
        ENDIF
        GOTO130
      ENDIF
C
      NUM = NUM + 1
      CLIQ(NUM) = ISUB2
      IF (CAND1(LEVEL1) .EQ. 0) THEN
C
C       Output a clique
C
        WRITE(*,160) (CLIQ(I),I=1,NUM)
```

```
160         FORMAT(' CLIQUE: ',20I3)
          ELSE
            IF (CAND2(LEVEL1) .LT.
     +        CAND1(LEVEL1)) THEN
              LEVEL = LEVEL + 1
              LEVEL1 = LEVEL1 + 1
              GOTO 40
            ENDIF
          ENDIF
C
170       NUM = NUM - 1
          CAND2(LEVEL) = CAND2(LEVEL) + 1
          IF (NDPS1(LEVEL) .GT. 1) THEN
            NODEW = CAND2(LEVEL)
C
C           Look for candidate
C
180         NODEW = NODEW + 1
            ITEMP = STACK(LEVEL,NODEW)
            IF (ITEMP .EQ. NDPS2(LEVEL))
     +        GOTO 180
            LOW = ARCFIR(ITEMP)
            IUP = ARCFIR(ITEMP + 1)
            IF (IUP .GT. LOW) THEN
              IUP = IUP - 1
              DO 190 K = LOW, IUP
                IF (FWDARC(K) .EQ. NDPS2(LEVEL))
     +            GOTO 180
190           CONTINUE
            ENDIF
          ENDIF
C
          NDPS1(LEVEL) = NDPS1(LEVEL) - 1
          GOTO 90
        ENDIF
C
        IF (LEVEL .GT. 1) THEN
          LEVEL = LEVEL - 1
```

```
        LEVEL1 = LEVEL1 - 1
        GOTO 170
      ENDIF
C
      RETURN
      END
```

2

SHORTEST PATHS

2-1 One-Pair Shortest Path

A. Problem description

Find the shortest path from a specified source node to another specified destination node in a directed graph with nonnegative edge lengths.

B. Method

As a means of reducing the computational requirements to find the shortest path from the source node s to the sink node t, the paths of both directions from s and into t will be considered. In general, all paths out of s and into t as far as their adjacent connected nodes are examined simultaneously. The path which has so far covered the least distance is extended. This process is repeated until a path is found out of s which has a node on it that already existed on a path into t, or vice versa. The complete path is then checked to see if the shortest path is obtained.

If n is the number of nodes in the graph, then the processing time of the entire algorithm is bounded by $O(n^2 \log n^2)$.

C. Subroutine SHORTP parameters

Input:

N	Number of nodes.
DIST	Integer matrix of dimension N by N; DIST(i, j) is the nonnegative length of the edge directed from node i to node j, with DIST(i, i) = 0 for all i.
NDIM	Row dimension of matrix DIST exactly as specified in the dimension statement of the calling program.
ISORCE	The specified source node.
ISINK	The specified destination node.
LARGE	A sufficiently large integer greater than the sum of all edge lengths in the graph.

Output:

NUMNOD	The number of nodes in the shortest path from ISORCE to ISINK.
PATH	Integer vector of length N; the nodes of the shortest path are stored in PATH(i), i = 1, 2, . . . , NUMNOD, where PATH(1) = ISORCE and PATH(NUMNOD) = ISINK
LNPATH	The length of the shortest path from ISORCE to ISINK.

Working storages:

POINT1	Integer vector of length N; POINT1(i) is the father of node i in the current path from ISORCE to node i.
POINT2	Integer vector of length N; POINT2(i) is the

	son of node i in the current path from node i to ISINK.
LEN1	Integer vector of length N; LEN1(i) is the length of the current path from ISORCE to node i.
LEN2	Integer vector of length N; LEN2(i) is the length of the current path from node i to ISINK.
STACK1	Integer vector of length N; stack of nodes corresponding to the array LEN1.
STACK2	Integer vector of length N; stack of nodes corresponding to the array LEN2.

D. Test example

Find the shortest path from node 3 to node 2 in the following graph.

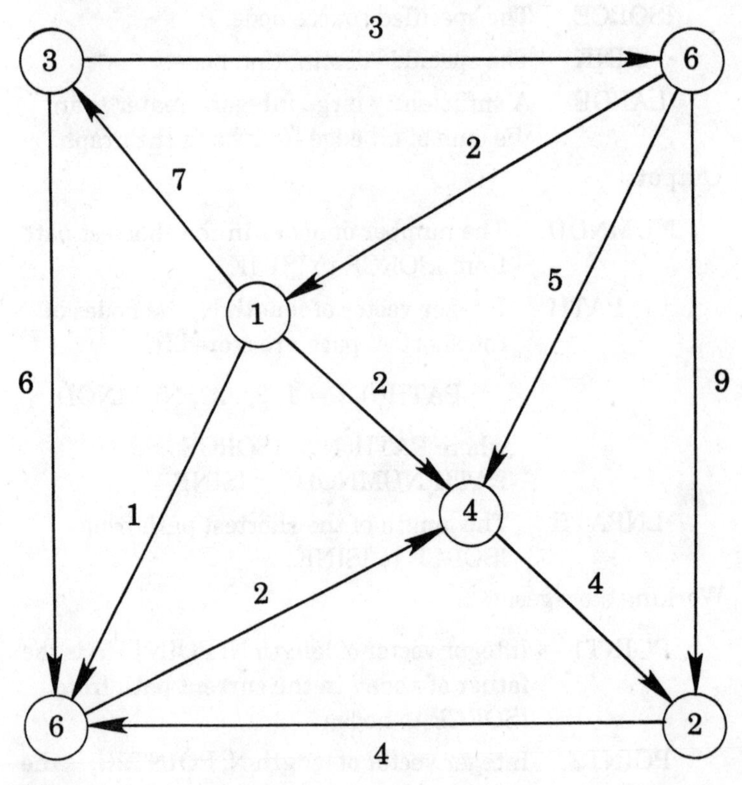

E. Program listing

MAIN PROGRAM

```
      INTEGER DIST(6,6),PATH(6),POINT1(6),
     +           POINT2(6),LEN1(6),LEN2(6),
     +           STACK1(6),STACK2(6)
      DATA DIST / 0,99,99,99,99,2, 99,0,99,4,99,9,
     +            7,99,0,99,99,99,  2,99,99,0,2,5,
     +            1,4,6,99,0,99,  99,99,3,99,0/
C
      N = 6
      NDIM = 6
      ISORCE = 3
      ISINK = 2
      LARGE = 999
      CALL SHORTP (N,DIST,NDIM,ISORCE,
     +              ISINK,LARGE,NUMNOD,
     +              PATH,LNPATH,POINT1,
     +              POINT2,LEN1,LEN2,
     +              STACK1,STACK2)
      WRITE(*,10) (PATH(I),I=1,NUMNOD)
   10 FORMAT(' THE SHORTEST PATH FOUND:',
     +        //1X,6I3)
      WRITE(*,20) LNPATH
   20 FORMAT(/' LENGTH OF THE SHORTEST',
     +        ' PATH = ',I4)
      STOP
      END
```

OUTPUT RESULTS

```
 THE SHORTEST PATH FOUND:
   3  6  1  4  2
 LENGTH OF THE SHORTEST PATH =  11
```

```
      SUBROUTINE SHORTP (N,DIST,NDIM,
     +                   ISORCE,ISINK,
```

```
     +                                LARGE, NUMNOD,
     +                                PATH,LNPATH,
     +                                POINT1,POINT2,
     +                                LEN1,LEN2,
     +                                STACK1,STACK2)
C
C     Find the shortest path from ISORCE to ISINK
C       in a directed graph
C
      INTEGER DIST(NDIM,N),PATH(N),
     +          POINT1(N),POINT2(N),
     +          LEN1(N),LEN2(N),STACK1(N),
     +          STACK2(N)
C
      KEY1 = 0
      KEY2 = 0
      DO 10 I = 1, N
        LEN1(I) = DIST(ISORCE,I)
        LEN2(I) = DIST(I,ISINK)
        POINT1(I) = ISORCE
        POINT2(I) = ISINK
   10 CONTINUE
C
C     Find the initial values of MINLN1 and
C       MINLN2 with corresponding INDEX values
C       for LEN1 and LEN2
C
      IPT1 = 0
      IPT2 = 0
      MINLN2 = LARGE
      MINLN1 = LARGE
      DO 20 I = 1, N
        CALL FMIN(N,LEN1,STACK1,KEY1,
     +    MINLN1,I,IPT1)
        CALL FMIN(N,LEN2,STACK2,KEY2,
     +    MINLN2,I,IPT2)
   20 CONTINUE
C
```

```
 30   IF (MINLN1 .LE. MINLN2) THEN
C
C       Reset LEN1
C
        KEY1 = MINLN1
 40     IF (IPT1 .GT. 0) THEN
          INDEX = STACK1(IPT1)
          DO 50 I = 1, N
            IDIST = DIST(INDEX,I)
            LENSUM = MINLN1 + IDIST
            IF (LEN1(I) .GT. LENSUM) THEN
              LEN1(I) = LENSUM
              POINT1(I) = INDEX
            ENDIF
 50       CONTINUE
          IPT1 = IPT1 - 1
          GOTO 40
        ENDIF
C
C       Find new MINLN1 and INDEX values for
C         LEN1
C
        MINLN1 = LARGE
        IPT1 = 0
        DO 60 I = 1, N
          CALL FMIN(N,LEN1,STACK1,KEY1,
     +      MINLN1,I,IPT1)
 60     CONTINUE
      ELSE
C
C       Reset LEN2
C
        KEY2 = MINLN2
 70     IF (IPT2 .GT. 0) THEN
          INDEX = STACK2(IPT2)
          DO 80 I = 1, N
            IDIST = DIST(I,INDEX)
            LENSUM = MINLN2 + IDIST
```

```
            IF (LEN2(I) .GT. LENSUM) THEN
              LEN2(I) = LENSUM
              POINT2(I) = INDEX
            ENDIF
 80       CONTINUE
          IPT2 = IPT2 - 1
          GOTO 70
        ENDIF
C
C       Find new MINLN2 and INDEX values for
C         LEN2
C
        MINLN2 = LARGE
        IPT2 = 0
        DO 90 I = 1, N
          CALL FMIN(N,LEN2,STACK2,KEY2,
     +      MINLN2,I,IPT2)
 90     CONTINUE
      ENDIF
C
C     Compute convergence criterion
C
      MINLN3 = LARGE
      DO 100 I = 1, N
        LENSUM = LEN1(I) + LEN2(I)
        IF (LENSUM .LT. MINLN3) THEN
          MINLN3 = LENSUM
          MNODE = I
        ENDIF
100   CONTINUE
      IF (MINLN3 .GT. MINLN1 + MINLN2)
     +  GOTO 30
C
C     Two ends of a shortest path meet in node
C       MNODE; unravel the path
C
      NUM1 = MNODE
      PATH(N) = MNODE
      IF (MNODE .NE. ISORCE) THEN
```

```
        NUMNOD = N - 1
110     NUM2 = POINT1(NUM1)
        IF (NUM2 .NE. ISORCE) THEN
          NUM1 = NUM2
          PATH(NUMNOD) = NUM2
          NUMNOD = NUMNOD - 1
          GOTO 110
        ENDIF
      ELSE
        NUMNOD = N
      ENDIF
      PATH(1) = ISORCE
      NUM1 = NUMNOD + 1
      NUMNOD = 2
120   IF (NUM1 .LE. N) THEN
        PATH(NUMNOD) = PATH(NUM1)
        NUMNOD = NUMNOD + 1
        NUM1 = NUM1 + 1
        GOTO 120
      ENDIF
      IF (MNODE .NE. ISINK) THEN
        NUM1 = MNODE
130     NUM2 = POINT2(NUM1)
        IF (NUM2 .NE. ISINK) THEN
          NUM1 = NUM2
          PATH(NUMNOD) = NUM2
          NUMNOD = NUMNOD + 1
          GOTO 130
        ENDIF
        PATH(NUMNOD) = ISINK
      ENDIF
C
      LNPATH = LEN1(MNODE) + LEN2(MNODE)
C
      RETURN
      END

      SUBROUTINE FMIN (N,DISTAB,STACK,
```

```
     +                      KEY,LITTLE,J,LEVEL)
C
C       Find LITTLE and store in STACK(1:LEVEL) all
C         values of k such that DISTAB(k) = LITTLE
C         and DISTAB(J) = KEY
C       (this subprogram is used by routine SHORTP)
C
        INTEGER DISTAB(N),STACK(N)
C
        DISTJ = DISTAB(J)
        IF (DISTJ .GT. KEY) THEN
          IF (DISTJ .LT. LITTLE) THEN
            LEVEL = 1
            LITTLE = DISTJ
            STACK(LEVEL) = J
          ELSE
            IF (DISTJ .EQ. LITTLE) THEN
              LEVEL = LEVEL + 1
              STACK(LEVEL) = J
            ENDIF
          ENDIF
        ENDIF
C
        RETURN
        END
```

2-2 One-to-All Shortest Path Lengths

A. Problem description

Find the lengths of all shortest paths from a specified source node to every other node in a given directed graph with non-negative edge lengths.

B. Method

Let $d(u, v)$ be the length of edge (u, v) in a complete directed graph of n nodes. The length of the shortest path from node 1 to node i—$S(i)$, $i = 2, 3, \ldots, n$—can be obtained as follows.

STEP 1. Set

$$p = 1, k = n, S(1) = 0, L(i) = i, S(i) = \infty$$

for $i = 2, 3, \ldots, n$.

STEP 2. For $i = 2, 3, \ldots, k$, do the following:

let $j = L(i)$ and compute $S(j) = \min(S(j), S(p) + d(p,j))$;

if the value of $S(j)$ is less than the current minimum, say $S(q)$, during this execution of Step 2, then set

$q = j$ and $t = i$.

STEP 3. Set

$$p = q, L(t) = L(k), k = k - 1.$$

If $k = 1$, then stop; otherwise, return to Step 2.

The algorithm requires $n^3/2$ additions and n^3 comparisons.

C. Subroutine SHOTLN parameters

Input:

N	Number of nodes.
DIST	Integer matrix of dimension N by N; DIST(i, j) is the nonnegative length of the edge directed from node i to node j, $\text{DIST}(i, i) = 0$ for all i.
NDIM	Row dimension of matrix DIST exactly as specified in the dimension statement of the calling program.
IROOT	The specified source mode.

LARGE	A sufficiently large integer greater than the sum of all edge lengths in the graph.

Output:

SHDIST	Integer vector of length N; SHDIST(i) is the length of the shortest path from IROOT to node i.

Working storages:

LOCATE	Integer vector of length N; array of node locations.

D. Test example

Find the lengths of the shortest paths from node 5 to every other node in the following graph.

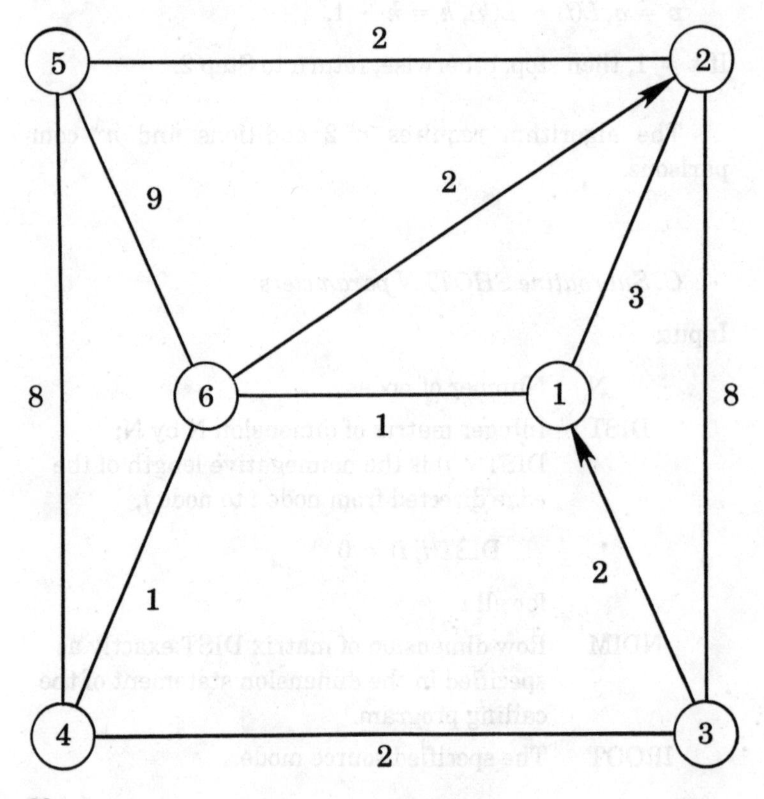

E. Program listing

MAIN PROGRAM

```
      INTEGER DIST(6,6),LOCATE(6),SHDIST(6)
      DATA DIST / 0,3,2,99,99,1, 3,0,8,99,2,2,
     +            99,8,0,2,99,99,
     +            99,99,2,0,8,1, 99,2,99,8,0,9,
     +            1,99,99,1,9,0/
      N = 6
      NDIM = 6
      IROOT = 5
      LARGE = 999
      CALL SHOTLN (N,DIST,NDIM,IROOT,LARGE,
        SHDIST,LOCATE)

      DO 20 I = 1, N
        IF (I .NE. IROOT) THEN
        WRITE(*,10) IROOT,I,SHDIST(I)
10      FORMAT(' THE SHORTEST DISTANCE',
     +         ' FROM NODE',I3,
     +         ' TO NODE',I3,' IS',I4)
        ENDIF
20    CONTINUE
      STOP
      END
```

OUTPUT RESULTS

```
THE SHORTEST DISTANCE FROM NODE 5
  TO NODE 1 IS 5
THE SHORTEST DISTANCE FROM NODE 5
  TO NODE 2 IS 2
THE SHORTEST DISTANCE FROM NODE 5
  TO NODE 3 IS 9
THE SHORTEST DISTANCE FROM NODE 5
  TO NODE 4 IS 7
THE SHORTEST DISTANCE FROM NODE 5
  TO NODE 6 IS 6
```

```
      SUBROUTINE SHOTLN (N,DIST,NDIM,
     +                           IROOT,LARGE,
     +                           SHDIST,LOCATE)
      INTEGER DIST(NDIM,N),SHDIST(N),
     +                 LOCATE(N)
C
C     Find the shortest distances from node IROOT to
C       all other nodes
C
      IF (IROOT .NE. 1) THEN
C
C       Interchange rows 1 and IROOT
C
        DO 10 I = 1, N
          ITEMP = DIST(1,I)
          DIST(1,I) = DIST(IROOT,I)
          DIST(IROOT,I) = ITEMP
 10     CONTINUE
C
C       Interchange columns 1 and IROOT
C
        DO 20 I = 1, N
          ITEMP = DIST(I,1)
          DIST(I,1) = DIST(I,IROOT)
          DIST(I,IROOT) = ITEMP
 20     CONTINUE
      ENDIF
C
      NODEU = 1
      N2 = N + 2
C
      DO 30 I = 1, N
        LOCATE(I) = I
        SHDIST(I) = DIST(NODEU,I)
 30   CONTINUE
C
      DO 50 I = 2, N
        K = N2 - I
        MINLEN = LARGE
```

```
          DO 40 J = 2, K
            NODEV = LOCATE(J)
            ITEMP = SHDIST(NODEU)
     +        + DIST(NODEU,NODEV)
            IF (ITEMP .LT. SHDIST(NODEV))
     +        SHDIST(NODEV) = ITEMP
            IF (MINLEN .GT. SHDIST(NODEV)) THEN
              MINLEN = SHDIST(NODEV)
              MINV = NODEV
              MINJ = J
            ENDIF
   40     CONTINUE
          NODEU = MINV
          LOCATE(MINJ) = LOCATE(K)
   50   CONTINUE
C
        IF (IROOT .NE. 1) THEN
          SHDIST(1) = SHDIST(IROOT)
          SHDIST(IROOT) = 0
C
C         Interchange rows 1 and IROOT
C
          DO 60 I = 1, N
            ITEMP = DIST(1,I)
            DIST(1,I) = DIST(IROOT,I)
            DIST(IROOT,I) = ITEMP
   60     CONTINUE
C
C         Interchange columns 1 and IROOT
C
          DO 70 I = 1, N
            ITEMP = DIST(I,1)
            DIST(I,1) = DIST(I,IROOT)
            DIST(I,IROOT) = ITEMP
   70     CONTINUE
        ENDIF
C
        RETURN
        END
```

2-3 One-to-All Shortest Path Tree

A. Problem description

Consider a directed graph with lengths associated with edges. The edge lengths may be nonpositive, but no cycles of negative lengths are present in the graph. The problem is to find the shortest paths from a given node to every other node.

B. Method

Let $d(u, v)$ be the length of the edge directed from node u to node v in the given graph. The shortest paths from a specified node s to every other node can be found by the following label-correcting method.

STEP 1. Let T be a tree consisting of the root s alone. Set

$$L(s) = 0 \text{ and } L(i) = \infty$$

for every other node i.

STEP 2. Find an edge (u, v) such that $L(u) + d(u, v) < L(v)$. Set

$$L(v) = L(u) + d(u, v).$$

Include the edge (u, v) in T and remove any edge in T which ends at v.

STEP 3. Repeat Step 2 until

$$L(u) + d(u, v) \geq L(v)$$

for all edges (u, v). The tree T will contain a shortest path from s to every other node.

The computation of the method is $O(n^3)$.

C. Subroutine SHTREE parameters

Input:

N	Number of nodes.
M	Number of edges.
N1	Equal to N + 1.
IROOT	The shortest paths from node IROOT to all other nodes are to be found.
INODE, JNODE	Each is an integer vector of length M; the *i*th edge is directed from INODE(*i*) to JNODE(*i*)
ARCLEN	Integer vector of length M; ARCLEN(*i*) is the length of the *i*th edge.
LARGE	A sufficiently large integer greater than the sum of all edge lengths in the graph.

Output:

DIST	Integer vector of length N; DIST(*i*) is the length of the shortest path from IROOT to node *i*; DIST(IROOT) is equal to zero.
TREARC	Integer vector of length N; the directed edge from node TREARC(*i*) to node *i* is in the shortest path tree; TREARC(IROOT) is equal to zero.

Working storages:

FWDARC	Integer vector of length M; FWDARC(*i*) is the ending node of the *i*th edge in the forward star representation of the graph.
ARCFIR	Integer vector of length N1; ARCFIR(*i*) is the number of the first edge starting at node *i* in the foward star representation of the graph.

ORIGIN	Integer vector of length M; ORIGIN(i) is the original edge number of the ith edge in the forward star representation of the graph.
MARK	Logical vector of length N; MARK(i) shows whether node i has been added to the list.
QUEUE	Integer vector of length N; first-in, first-out queue for the examination of the nodes.

D. Test example

Find the shortest paths from node 3 to all other nodes in the following graph.

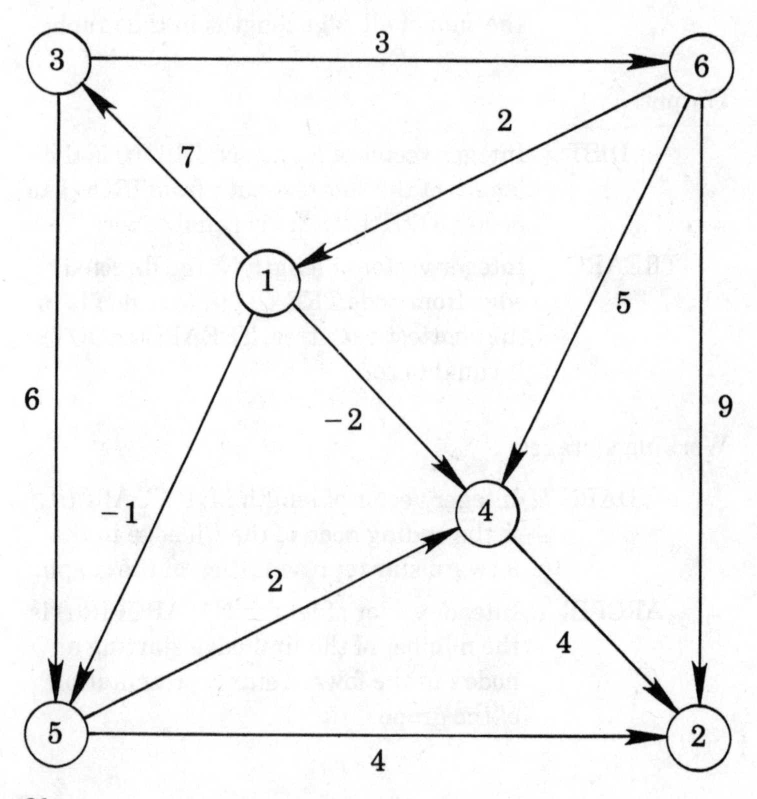

E. Program listing

MAIN PROGRAM

```
      INTEGER INODE(11),JNODE(11),
     +        ARCLEN(11),DIST(6),TREARC(6),
     +        FWDARC(11),ARCFIR(7),
     +        ORIGIN(11),QUEUE(6)
      LOGICAL MARK(6)
      DATA INODE /3, 2, 5, 4, 1, 1, 6, 6, 1, 3, 6/,
     +     JNODE /6, 5, 4, 2, 4, 5, 4, 1, 3, 5, 2/,
     +     ARCLEN / 3, 4, 2, 4, -2, -1, 5, 2, 7, 6, 9/
C
      N = 6
      M = 11
      IROOT = 3
      N1 = N + 1
      LARGE = 999
      CALL SHTREE (N,M,N1,IROOT,INODE,
     +             JNODE,ARCLEN,LARGE,
     +             DIST,TREARC,FWDARC,
     +             ARCFIR,ORIGIN,MARK,
     +             QUEUE)
      DO 20 I = 1, N
        IF (I .NE. IROOT) THEN
          WRITE(*,10) IROOT,I,DIST(I)
10        FORMAT(' THE SHORTEST PATH',
     +           ' DISTANCE FROM NODE',I3,
     +           ' TO NODE',I3,' IS',I4)
        ENDIF
20    CONTINUE
      WRITE(*,30)
30    FORMAT(/' THE EDGES IN THE SHORTEST',
     +        ' PATH TREE ARE:')
      DO 50 I = 1, N
        IF (I .NE. IROOT) THEN
          WRITE(*,40) TREARC(I),I
40        FORMAT(15X,'(',I1,',',I1,')')
        ENDIF
```

```
50    CONTINUE
      STOP
      END
```

OUTPUT RESULTS

```
THE SHORTEST PATH DISTANCE FROM NODE 3
  TO NODE 1 IS 5
THE SHORTEST PATH DISTANCE FROM NODE 3
  TO NODE 2 IS 7
THE SHORTEST PATH DISTANCE FROM NODE 3
  TO NODE 4 IS 3
THE SHORTEST PATH DISTANCE FROM NODE 3
  TO NODE 5 IS 4
THE SHORTEST PATH DISTANCE FROM NODE 3
  TO NODE 6 IS 3
THE EDGES IN THE SHORTEST PATH TREE ARE:
                 (6,1)
                 (4,2)
                 (1,4)
                 (1,5)
                 (3,6)
```

```
      SUBROUTINE SHTREE (N,M,N1,IROOT,
     +                   INODE,JNODE,
     +                   ARCLEN,LARGE,
     +                   DIST,TREARC,
     +                   FWDARC,ARCFIR,
     +                   ORIGIN,MARK,
     +                   QUEUE)
C
C     Find the shortest paths from IROOT to all other
C       nodes
C
      INTEGER INODE(M),JNODE(M),ARCLEN(M),
     +        DIST(N),TREARC(N),
     +        FWDARC(M),ARCFIR(N1),
     +        ORIGIN(M),QUEUE(N)
      LOGICAL MARK(N)
C
```

```
C      Set up the forward star representation of the
C         graph
C
       K = 0
       DO 20 I = 1, N
          ARCFIR(I) = K + 1
          DO 10 J = 1, M
             IF (INODE(J) .EQ. I) THEN
                K = K + 1
                ORIGIN(K) = J
                FWDARC(K) = JNODE(J)
             ENDIF
  10      CONTINUE
  20   CONTINUE
       ARCFIR(N1) = M + 1
C
       DO 40 I = 1, N
          TREARC(I) = 0
          MARK(I) .TRUE.
          DIST(I) = LARGE
  40   CONTINUE
       DIST(IROOT) = 0
       NODEV = 1
       NODEY = NODEV
       NODEU = IROOT
C
  50   LENU = DIST(NODEU)
       ISTART = ARCFIR(NODEU)
       IF (ISTART .NE. 0) THEN
          INDEX = NODEU + 1
  60      LAST = ARCFIR(INDEX) - 1
          IF (LAST .LE. -1) THEN
             INDEX = INDEX + 1
             GOTO 60
          ENDIF
       DO 70 I = ISTART, LAST
             IWW = FWDARC(I)
             LENSUM = ARCLEN(ORIGIN(I)) + LENU
             IF (DIST(IWW) .GT. LENSUM) THEN
```

```
                DIST(IWW) = LENSUM
                TREARC(IWW) = NODEU
                IF (MARK(IWW)) THEN
                  MARK(IWW) = .FALSE.
                  QUEUE(NODEY) = IWW
                  NODEY = NODEY + 1
                  IF (NODEY .GT. N) NODEY = 1
                ENDIF
              ENDIF
   70       CONTINUE
          ENDIF
          IF (NODEV .NE. NODEY) THEN
            NODEU = QUEUE(NODEV)
            MARK(NODEU) = .TRUE.
            NODEV = NODEV + 1
            IF (NODEV .GT. N) NODEV = 1
            GOTO 50
          ENDIF
C
          RETURN
          END
```

2-4 All-Pairs Shortest Paths

A. Problem description

Find the shortest paths between all pairs of nodes in a directed graph with given edge lengths, assuming that there is no cycle of negative length in the graph.

B. Method

Let L be a sufficiently large integer. Initially set $d(i,j)$ equal to the length of the edge from node i to node j in the given directed graph of n nodes. It is assumed that $d(i, i) = 0$, for $i = 1, 2, \ldots, n$, and $d(i, j) = L$ if the edge (i, j) is not in the graph.

STEP 1. Set $k = 0$.

STEP 2. Set $k = k + 1$.

STEP 3. For all $i \neq k$, such that $d(i, k) \neq L$, and all $j \neq k$, such that $d(k, j) \neq L$, compute

$$d(i, j) = \min(d(i, j), d(i, k) + d(k, j)).$$

STEP 4. If $k = n$ then stop; $d(i, j)$ are the lengths of all the shortest paths for all i and j; otherwise, return to Step 2.

The running time of the algorithm is $O(n^3)$.

C. Subroutine ALLPTH parameters

Input:

N	Number of nodes.
DIST	Real matrix of dimension N by N; DIST(i, j) is the length of the edge directed from node i to node j, DIST(i, i) = 0 for all i, and DIST(i, j) = BIG if edge (i, j) is not in the graph. On ouput, DIST(i, j) is the length of the shortest path from node i to node j.
NDIM	Row dimension of matrix DIST exactly as specified in the dimension statement of the calling program.
BIG	A sufficiently large real number greater than the sum of all edge lengths in the graph.

Output:

NEXT	Integer matrix of dimension N by N; NEXT(i, j) is the next-to-last node in the shortest path from node i to node j. This array can be used to trace the shortest path for every pair of nodes.

D. Test example

Find the lengths of the shortest paths between all pairs of nodes in the following graph. Furthermore, output the shortest path from node 5 to node 3.

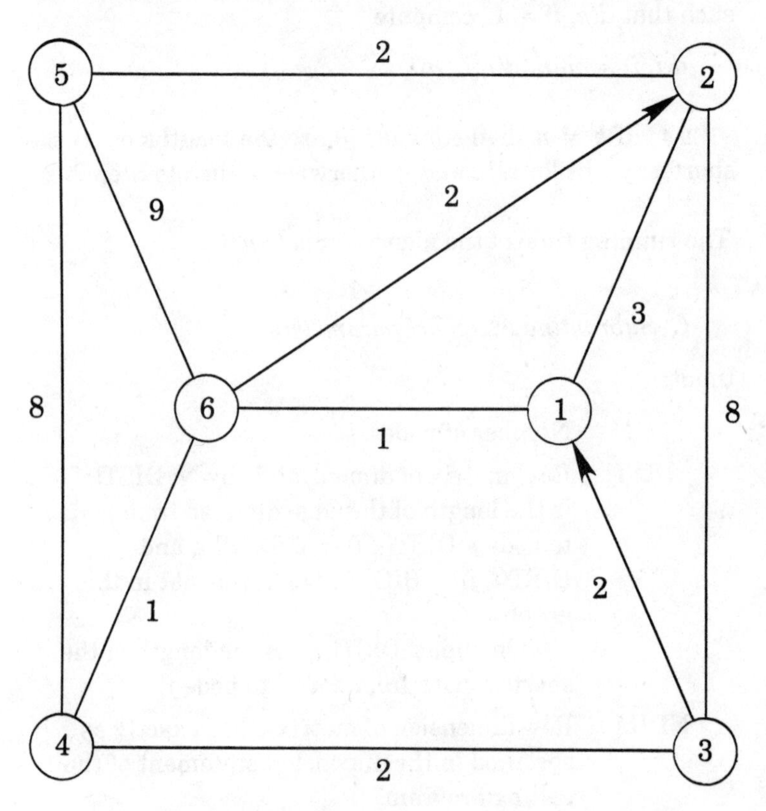

E. Program listing

MAIN PROGRAM

```
      INTEGER NEXT(6,6),PATH(6)
      REAL    DIST(6,6)
      DATA DIST / 0.,3.,2.,99.,99.,1., 3.,0.,8.,99.,2.,2.,
     +            99.,8.,0.,2.,99.,99., 99.,99.,2.,0.,8.,1.,
     +            99.,2.,99.,8.,0.,9., 1.,99.,99.,1.,9.,0./
      N = 6
      NDIM = 6
```

```
      BIG = 999.
      CALL ALLPTH (N,DIST,NDIM,BIG,NEXT)
      WRITE(*,10)
10    FORMAT(' SHORTEST DISTANCE MATRIX',
     +       ' BETWEEN ALL NODE PAIRS'/)
      DO 20 I = 1, N
20       WRITE (*,30) (DIST(I,J),J=1,N)
30    FORMAT(4X,6F5.0)
C
C     Store the shortest path from node 5 to node 3
C        in array PATH
C
      NODE1 = 5
      NODE2 = 3
      J = NODE2
      NUM = 1
      PATH(NUM) = NODE2
40    NODE = NEXT(NODE1,J)
      NUM = NUM + 1
      PATH(NUM) = NODE
      IF (NODE .NE. NODE1) THEN
         J = NODE
         GOTO 40
      ENDIF
      WRITE(*,50) NODE1,NODE2,(PATH(I),
     +   I=1,NUM)
50    FORMAT(/' THE SHORTEST PATH FROM',
     +       ' NODE',I3' TO NODE ',
     +       I3,':'/
     +   ' (IN REVERSE ORDER)'//2X,6I4)
      STOP
      END
```

OUTPUT RESULTS

```
SHORTEST DISTANCE MATRIX BETWEEN ALL
  NODE PAIRS

                0. 3. 4. 2. 5. 1.
                3. 0. 7. 5. 2. 4.
```

```
                2. 5. 0. 2. 7. 3.
                2. 3. 2. 0. 5. 1.
                5. 2. 9. 7. 0. 6.
                1. 2. 3. 1. 4. 0.
THE SHORTEST PATH FROM NODE 5 to NODE 3:
  (IN REVERSE ORDER)
                  3 4 6 1 2 5
```

```
      SUBROUTINE ALLPTH (N,DIST,NDIM,BIG,
     +                       NEXT)
C
C     Find the shortest paths for all pairs of nodes
C
      INTEGER NEXT(N,N)
      REAL    DIST(NDIM,N)
C
      DO 10 I = 1, N
        DO 10 J = 1, N
 10       NEXT(I,J) = I
C
      DO 40 I = 1, N
        DO 30 J = 1, N
          IF (DIST(J,I) .LT. BIG) THEN
            DO 20 K = 1, N
              IF (DIST(I,K) .LT. BIG) THEN
                D = DIST(J,I) + DIST(I,K)
                IF (D .LT. DIST(J,K)) THEN
                  DIST(J,K) = D
                  NEXT(J,K) = NEXT(I,K)
                ENDIF
              ENDIF
 20         CONTINUE
          ENDIF
 30     CONTINUE
 40   CONTINUE
C
      RETURN
      END
```

2-5 k Shortest Paths

A. Problem description

Consider a directed graph of n nodes with lengths on edges. It is assumed that all cycles in the graph have strictly positive lengths. Given an integer $k > 1$ and a specified node s, the problem is to find k shortest distinct path lengths from node s to every other node in the graph. Furthermore, list the k shortest paths from node s to every other node. As defined here, the k shortest paths are allowed to contain embedded cycles, that is, some nodes may be repeated in a path.

B. Method

The k shortest path problem in a graph of n nodes with given edge lengths $d(i,j)$ will be solved by a label-correcting method in which an initial guess is given to the k shortest path lengths. Then the tentative k shortest path lengths will be improved successively.

A sequence of alternating forward and backward iterations will be employed. During the forward iteration, the nodes are examined in the order $1, 2, \ldots, n$, and only edges (i,j) with $i < j$ are processed. During the backward iteration, the nodes are examined in the order $n, n-1, \ldots, 1$, and only edges (i, j) with $i > j$ are processed. The alternating forward and backward iterations are continued until the node labels at two consecutive iterations coincide, in which case no improvement is possible.

The procedure of finding the k shortest path lengths from node 1 to all other nodes can be outlined as follows.

STEP 1. Initialize the required k shortest path lengths, a k-vector

$$S(i) = (s(i, 1), s(i, 2), \ldots, s(i, k)),$$

from node 1 to node i by setting

$$S(1) = (0, \infty, \ldots, \infty),$$

and

$$S(i) = (\infty, \infty, \ldots, \infty), \text{ for } i = 2, 3, \ldots, n.$$

STEP 2. For $j = 2$ to n, do the following: For every node i adjacent to node j, where $i < j$, if

$$\{s(i, p) + d(i, j): p = 1, 2, \ldots, k\}$$

gives a smaller path length than any one of the tentative k shortest path lengths in $S(j)$, then the current k-vector $S(j)$ is updated by inclusion of this smaller path length.

STEP 3. If none of the node labels $S(j)$ has changed in Step 2, then stop.

STEP 4. For $j = n - 1$ to 1, do the following: For every node i adjacent to node j, where $i > j$, process the edge (i, j) the same way as in Step 2.

STEP 5. If none of the node labels $S(j)$ has changed in Step 4, then stop; otherwise, return to Step 2.

After the k shortest distinct path lengths are determined by the above procedure, the paths themselves can be reconstructed from the path length information. In essence, if a jth shortest path P of length t from node u to node v passes through node w, then the subpath of P extending from node u to node w is an ith shortest path, for some i, $1 \leq i \leq j$. This can be used to determine the penultimate node w on a jth shortest path of known length t from node u to node v. This backtracking method can be applied repeatedly to produce all the k shortest paths from node u to node v.

It should be noted that several paths can have the same

path lengths. It is therefore possible that more than k paths may be generated from the k distinct shortest path lengths.

The whole algorithm will take at most $O(n^2k^2\log n)$ operations.

The subroutine KSHOT1 below will calculate the k shortest path lengths and subroutine PRTPTH will generate the shortest paths.

C. Subroutine KSHOT1 parameters

Input:

N	Number of nodes.
M	Number of edges.
NIJ	Number of edges (i, j) in which $i > j$.
NJI	Number of edges (i, j) in which $j > i$.
K	Number of shortest paths.
ISORCE	The K shortest paths from node ISORCE to all other nodes are to be found.
ITER	IF ITER = 0 then the maximum number of iterations is set to 100; if ITER > 0 then it denotes the maximum number of iterations allowed. On output, ITER is the total number of iterations actually performed.
INODE, JNODE	Each is an integer vector of length M; the ith edge is directed from INODE(i) to JNODE(i), and it is assumed that the array JNODE is already sorted in non-decreasing order.
ARCLEN	Integer vector of length M; ARCLEN(i) is the length of the ith edge.

Output:

DIST Integer matrix of dimension N by K; DIST(i, j)

is the length of the jth shortest path from ISORCE to node i.

If node i is not reachable from ISORCE in the jth shortest path, then DIST(i, j) will be set equal to a number greater than the sum of all edge lengths in the input graph.

Working storages:

IEDGE1	Integer vector of length N; IEDGE1(i) is the number of edges (j, i), where $i < j$.
INOD1	Integer vector of length NIJ; array containing the nodes i incident to node j, where $i > j$ in the order of increasing j.
ILEN1	Integer vector of length NIJ; array of edge lengths corresponding to edges in INOD1.
IEDGE2	Integer vector of length N; IEDGE2(i) is the number of edges (j, i), where $j < i$.
INOD2	Integer vector of length NJI; array containing the nodes i incident to node j, where $i < j$ in the order of increasing j.
ILEN2	Integer vector of length NJI; array of edge lengths corresponding to edges in INOD2.
IAUX	Integer vector of length K; auxiliary storage for the K smallest numbers of an array.

Subroutine PRTPTH parameters

Input:

N	Number of nodes.
M	Number of edges.
N1	Equal to N + 1.
NPMAX	Maximum number of nodes in a possible path which allows repeated nodes.

K	Number of shortest paths.
MAXPTH	Maximum number of paths to be generated.
ISORCE ISINK	The K shortest paths from node ISORCE to node ISINK are to be generated.
INODE, JNODE, ARCLEN	See the parameter description in subroutine KSHOT1.
DIST	Output from subroutine KSHOT1.

Working storages:

FIRARC	Integer vector of length N1; contains the nodes on a path from ISORCE to ISINK.
NODINC	Integer vector of length M; array of nodes i incident to node j in the order of increasing j.
LENINC	Integer vector of length M; array of edge lengths corresponding to NODINC.
PATHND	Integer vector of length NPMAX: contains the nodes on a path from ISORCE to ISINK.
POINT	Integer vector of length NPMAX; POINT(i) is the position of node PATHND(i).
PLEN	Integer vector of length NPMAX; PLEN(i) is the edge length from node PATHND(i) to node PATHND($i - 1$).

D. Test example

Find the four shortest distinct path lengths from node 2 to every other node, and output at most 10 of the shortet paths from node 2 to node 5 in the following graph.

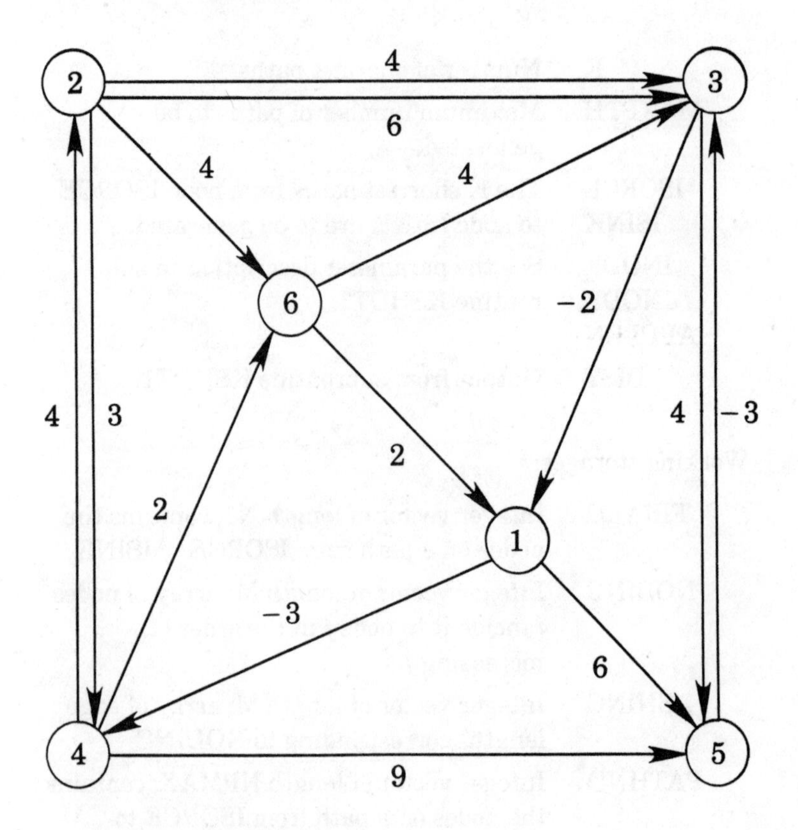

E. Program listing

MAIN PROGRAM

```
      INTEGER INODE(14),JNODE(14),
     +        ARCLEN(14),DIST(6,4),
     +        IEDGE1(6),INOD1(5),ILEN1(5),
     +        IEDGE2(6),INOD2(9),
     +        ILEN2(9),IAUX(4),
     +        FIRARC(7),NODINC(14),
     +        LENINC(14),
     +        PATHND(20),POINT(20),PLEN(20)
      DATA INODE /6, 3, 4, 5, 6, 2, 2, 2, 1, 4, 1, 3, 2, 4/,
     +     JNODE /1, 1, 2, 3, 3, 3, 3, 4, 4, 5, 5, 5, 6, 6/
     +     ARCLEN /2, -2, 3, -3, 4, 6, 4, 4, -3, 9,
     +             6, 4, 4, 2/
C
```

```
      N = 6
      M = 14
      NIJ = 5
      NJI = 9
      ISORCE = 2
      ITER = 0
      K = 4
C
C     Find the k shortest distinct path lengths
C
      CALL KSHOT1 (N,M,NIJ,NJI,K,ISORCE,
     +                    ITER,INODE,JNODE,
     +                    ARCLEN,DIST,IEDGE1,
     +                    INOD1,ILEN1,IEDGE2,
     +                    INOD2,ILEN2,IAUX)
C
      WRITE(*,30) K
 30   FORMAT (10X,' K =',I3,
     +   ' SHORTEST PATH LENGTHS:'/)
      DO 40 I = 1, N
 40     WRITE (*,50) ISORCE,I,(DIST(I,J),J=1,K)
 50   FORMAT(' FROM INODE',I3,
     +   ' TO NODE',I3,':',6X,(10I6))
      WRITE(*,60) ITER
 60   FORMAT(/' TOTAL NUMBER OF',
     +   ' ITERATIONS =',I4//)
C
C     Generate at most 10 of the shortest paths from
C       ISORCE to ISINK
C
      N1 = N + 1
      NPMAX = 20
      ISINK = 5
      MAXPTH = 10
      CALL PRTPTH (N,M,N1,NPMAX,K,
     +                    MAXPTH,ISORCE,ISINK,
     +                    INODE,JNODE,ARCLEN,
     +                    DIST,FIRARC,
     +                    NODINC,LENINC,
```

```
     +                    PATHND,POINT,PLEN)
C
      STOP
      END
```

OUTPUT RESULTS

```
K = 4 SHORTEST PATH LENGTHS:
FROM NODE 2 TO NODE 1:   2 3  4  5
FROM NODE 2 TO NODE 2:   0 2  3  4
FROM NODE 2 TO NODE 3:   4 5  6  7
FROM NODE 2 TO NODE 4:  -1 0  1  2
FROM NODE 2 TO NODE 5:   8 9 10 11
FROM NODE 2 TO NODE 6:   1 2  3  4
TOTAL NUMBER OF ITERATIONS = 6
THE FOLLOWING ARE THE SHORTEST PATHS
  FROM NODE 2 TO NODE 5
       LENGTH    N IN THE PATH
  1:     8       2 3 1 4 5
  2:     8       2 3 1 5
  3:     8       2 3 5
  4:     9       2 3 1 4 6 1 4 5
  5:     9       2 3 1 4 5 3 1 4 5
  6:     9       2 3 1 5 3 1 4 5
  7:     9       2 3 5 3 1 4 5
  8:     9       2 3 1 4 6 3 1 4 5
  9:     9       2 3 1 4 6 1 5
 10:     9       2 3 1 4 5 3 1 5
```

```
      SUBROUTINE KSHOT1 (N,M,NIJ,NJI,K,
     +                   ISORCE,ITER,
     +                   INODE,JNODE,
     +                   ARCLEN, DIST,
     +                   IEDGE1,INOD1,
     +                   ILEN1,IEDGE2,
     +                   INOD2,ILEN2,IAUX)
C
C     Find the K shortest distinct path lengths
```

```
C          allowing repeated nodes
C
        INTEGER INODE(M),JNODE(M),ARCLEN(M),
     +           DIST(N,K),IEDGE1(N),
     +           INOD1(NIJ),ILEN1(NIJ),
     +           IEDGE2(N),INOD2(NJI),
     +           ILEN2(NJI),IAUX(K)
        LOGICAL BEST
C
        MAXITR = ITER
        IF (ITER .LE. 0) MAXITR = 100
        INIT = 0
        NJI1 = 0
        NIJ1 = 0
        INDEX = 0
        LARGE = 1
        DO 20 I = 1, M
          LARGE = LARGE + ARCLEN(I)
          NODEV = INODE(I)
          NODEU = JNODE(I)
          LEN = ARCLEN(I)
          IF (NODEU .NE. INDEX) THEN
            IF (NODEU .NE. INDEX+1) THEN
              J1 = INDEX + 1
              J2 = NODEU - 1
              DO 10 J = J1, J2
                IEDGE2(J) = 0
   10           IEDGE1(J) = 0
            ENDIF
            IF (INIT .NE. 0) THEN
              IEDGE1(INDEX) = ISUB1
              IEDGE2(INDEX) = ISUB2
            ENDIF
            ISUB1 = 0
            ISUB2 = 0
            INDEX = NODEU
          ENDIF
          INIT = INIT + 1
```

```
        IF (NODEV .LE. NODEU) THEN
          NJI1 = NJI1 + 1
          INOD2(NJI1) = NODEV
          ILEN2(NJI1) = LEN
          ISUB2 = ISUB2 + 1
        ELSE
          NIJ1 = NIJ1 + 1
          INOD1(NIJ1) = NODEV
          ILEN1(NIJ1) = LEN
          ISUB1 = ISUB1 + 1
        ENDIF
 20   CONTINUE
      IEDGE2(INDEX) = ISUB2
      IEDGE1(INDEX) = ISUB1
C
      DO 30 I = 1, N
        DO 30 J = 1, K
 30       DIST(I,J) = LARGE
      DIST(ISORCE,1) = 0
      ITER = 1
C
 40   IDS2 = NIJ1
      BEST = .TRUE.
      I = N - 1
 50   IF (I .GT. 0) THEN
        IF (IEDGE1(I) .NE. 0) THEN
          IDS1 = IDS2 - IEDGE1(I) + 1
C
C         Matrix multiplication with DIST using the
C           lower triangular part of the edge distance
C           matrix
C
          DO 60 J = 1, K
 60         IAUX(J) = DIST(I,J)
          MAX = IAUX(K)
          DO 100 LOOP 1 = IDS1,IDS2
            INDEX 1 = INOD1(LOOP1)
            IDIST3 = ILEN1(LOOP1)
            DO 90 LOOP2 = 1, K
```

```
                IDIST1 = DIST(INDEX1,LOOP2)
                IF (IDIST1 .GE. LARGE) GOTO 100
                IDIST2 = IDIST1 + IDIST3
                IF (IDIST2 .GE. MAX) GOTO 100
                J = K
 70             IF (J .GE. 2) THEN
                  IF (IDIST2 .LT. IAUX(J-1)) THEN
                    J = J - 1
                    GOTO 70
                  ENDIF
                  IF (IDIST2 .EQ. IAUX(J-1))
     +              GOTO 90
                ELSE
                  J = 1
                ENDIF
                INDEX2 = K
 80             IF (INDEX2 .GT. J) THEN
                  IAUX(INDEX2)
     +              = IAUX(INDEX2 - 1)
                  INDEX2 = INDEX2 - 1
                  GOTO 80
                ENDIF
                IAUX(J) = IDIST2
                BEST = .FALSE.
                MAX = IAUX(K)
 90           CONTINUE
 100        CONTINUE
C
            IF (.NOT. BEST) THEN
              DO 110 J = 1, K
 110            DIST(I,J) = IAUX(J)
            ENDIF
            IDS2 = IDS1 - 1
          ENDIF
          I = I - 1
          GOTO 50
        ENDIF
        IF (ITER .NE. 1) THEN
          IF (BEST) RETURN
```

```
        ENDIF
        ITER = ITER + 1
        IDS1 = 1
        BEST = .TRUE.
        DO 180 I = 2, N
          IF (IEDGE2(I) .NE. 0) THEN
            IDS2 = IDS1 + IEDGE2(I) - 1
C
C           Matrix multiplication with DIST using the
C           upper triangular part of the edge distance
C           matrix
C
            DO 120 J = 1, K
 120          IAUX(J) = DIST(I,J)
            MAX = IAUX(K)
            DO 160 LOOP1 = IDS1,IDS2
              INDEX1 = INOD2(LOOP1)
              IDIST3 = ILEN2(LOOP1)
              DO 150 LOOP2 = 1, K
                IDIST1 = DIST(INDEX1,LOOP2)
                IF (IDIST1 .GE. LARGE) GOTO 160
                IDIST2 = IDIST1 + IDIST3
                IF (IDIST2 .GE. MAX) GOTO 160
                J = K
 130            IF (J .GE. 2) THEN
                  IF (IDIST2 .LT. IAUX(J - 1)) THEN
                    J = J - 1
                    GOTO 130
                  ENDIF
                  IF (IDIST2 .EQ. IAUX(J - 1))
     +              GOTO 150
                ELSE
                  J = 1
                ENDIF
                INDEX2 = K
 140            IF (INDEX2 .GT. J) THEN
                  IAUX(INDEX2)
```

```
     +               = IAUX(INDEX2 - 1)
                   INDEX2 = INDEX2 - 1
                   GOTO 140
                 ENDIF
                 IAUX(J) = IDIST2
                 BEST = .FALSE.
                 MAX = IAUX(K)
 150           CONTINUE
 160         CONTINUE
C
             IF (.NOT. BEST) THEN
               DO 170 J = 1, K
 170             DIST(I,J) = IAUX(J)
             ENDIF
             IDS1 = IDS2 + 1
           ENDIF
 180     CONTINUE
C
         IF (.NOT. BEST) THEN
           ITER = ITER + 1
           IF (ITER .LT. MAXITR) GOTO 40
         ENDIF
C
         RETURN
         END

         SUBROUTINE PRTPTH (N,M,N1,NPMAX,
     +                      K,MAXPTH,
     +                      ISORCE,ISINK,
     +                      INODE,JNODE,
     +                      ARCLEN,DIST,
     +                      FIRARC,NODINC,
     +                      LENINC,PATHND,
     +                      POINT,PLEN)
C
C        Output at most MAXPTH number of K shortest
C          paths from ISORCE to ISINK, allowing
```

```
C        repeated nodes (this subprogram is used after
C        subroutine KSHOT1)
C
      INTEGER INODE(M),JNODE(M),
     +        ARCLEN(M),DIST(N,K),
     +        FIRARC(N1),NODINC(M),
     +        LENINC(M),
     +        PATHND(NPMAX),
     +        POINT(NPMAX),PLEN(NPMAX)
C
      INIT = 0
      INDEX = 0
      LARGE = 1
C
      DO 20 I = 1, M
        LARGE = LARGE + ARCLEN(I)
        NODEV = INODE(I)
        NODEU = JNODE(I)
        LEN = ARCLEN(I)
        IF (NODEU .NE. INDEX) THEN
          IF (NODEU .NE. INDEX+1) THEN
            J1 = INDEX + 1
            J2 = NODEU - 1
            DO 10 J = J1, J2
   10         FIRARC(J) = 0
          ENDIF
          FIRARC(NODEU) = INIT + 1
          INDEX = NODEU
        ENDIF
        INIT = INIT + 1
        NODINC(INIT) = NODEV
        LENINC(INIT) = LEN
   20 CONTINUE
      FIRARC(INDEX + 1) = INIT + 1
C
      DO 30 I = 1, NPMAX
        PATHND(I) = 0
        POINT(I) = 0
```

```
         PLEN(I) = 0
   30   CONTINUE
        IPATH = 1
        IF (ISORCE .EQ. ISINK) IPATH = 2
        NUMPTH = 0
        IF (DIST(ISINK,IPATH) .GE. LARGE) THEN
          WRITE(*,40) ISORCE,ISINK
   40     FORMAT(' THERE IS NO PATH FROM',
     +                'NODE',I4,' TO NODE',I4)
          RETURN
        ENDIF
C
        WRITE(*,50) ISORCE,ISINK
   50   FORMAT(//' THE FOLLOWING ARE THE',
     +  ' SHORTEST PATHS FROM NODE',I3,
     +          'TONODE',I3//9X,'LENGTH',8X,'N',
     +          'IN THE PATH'/)
C
   60   IPTAG = 1
        IDST1 = DIST(ISINK,IPATH)
        IF (IDST1 .EQ. LARGE) RETURN
        IDST2 = IDST1
        PATHND(1) = ISINK
C
   70   IPT1 = 0
   80   NT = PATHND(IPTAG)
        IDS1 = FIRARC(NT)
        ND = NT
   90   IF ((FIRARC(ND + 1) .EQ. 0) .AND.
     +    (ND .LT. N)) THEN
          ND = ND + 1
          GOTO 90
        ENDIF
        IF = FIRARC(ND + 1) - 1
        IPT2 = IDS1 + IPT1
C
  100   IF (IPT2 .LE. IF) THEN
          ISUB = NODINC(IPT2)
```

```
      JLEN = LENINC(IPT2)
      LT = IDST1 - JLEN
C
      J = 1
110   IF ((DIST(ISUB,J) .GT. LT) .OR.
     +   (J .GT. K)) THEN
        IPT2 = IPT2 + 1
        GOTO 100
      ENDIF
      IF (DIST(ISUB,J) .LT. LT) THEN
        J = J + 1
        GOTO 110
      ENDIF
      IPTAG = IPTAG + 1
      IF (IPTAG .GT. NPMAX) THEN
        WRITE(*,120) NPMAX
120     FORMAT(' NUMBER OF ARCS IN A',
     +         ' PATH EXCEEDS',I6/
     +         ' REMEDY: INCREASE THE',
     +         ' VALUE OF NPMAX')
        RETURN
      ENDIF
C
      PATHND(IPTAG) = ISUB
      POINT(IPTAG) = IPT2 - IDS1 + 1
      PLEN(IPTAG) = JLEN
      IDST1 = LT
C
      IF (IDST1 .NE. 0) GOTO 70
      IF (ISUB .NE. ISORCE) GOTO 70
C
      NUMPTH = NUMPTH + 1
      WRITE(*,130) NUMPTH,IDST2,
     +   (PATHND(IPTAG-J+1),J=1,IPTAG)
130   FORMAT(1X,I3,':',I8,5X,(20I5))
      IF (NUMPTH .GE. MAXPTH) RETURN
    ENDIF
C
    IPT1 = POINT(IPTAG)
```

```
      PATHND(IPTAG) = 0
      IDST1 = IDST1 + PLEN(IPTAG)
      IPTAG = IPTAG - 1
      IF (IPTAG .GT. 0) GOTO 80
C
      IPATH = IPATH + 1
      IF (IPATH .LE. K) GOTO 60
C
      RETURN
      END
```

2-6 *k* Shortest Paths Without Repeated Nodes

A. Problem description

Consider a directed graph of n nodes with lengths on edges and which has no cycle of negative length. Given an integer $k > 0$, a specified source node s and a specified sink node t, the problem is to find the k shortest paths from s to t such that each of the k shortest paths does not have any embedded cycle. That is, each path does not contain any repeated nodes.

B. Method

Let $P(j)$ be the jth shortest path from the source s to the sink t. Among the shortest paths $P(1), P(2), \ldots, P(j-1)$ that have the same initial subpaths from node s to the ith node, let $Q(i, j)$ be the shortest of the paths that coincide with $P(j-1)$ from node s up to the ith node and then deviate to a node that is different from any $(i+1)$th node in other paths. The procedure of finding the k shortest paths from s to t follows.

STEP 1. Find $P(1)$, a shortest path from s to t. If there is only one shortest path, then enter it into a list $L1$. If there is more than one shortest path, then enter one into a list $L1$, and the rest into another list $L2$. Set $j = 2$.

STEP 2. For each $i = 1, 2, \ldots, j - 1$, find all $Q(i,j)$ and place them into the list $L2$.

STEP 3. Take a path from the end of the list $L2$. Denote this path by $P(j)$ and move it from $L2$ to $L1$.

If $j = k$ then stop—the required list of k shortest paths is in $L1$; otherwise, set $j = j + 1$ and return to Step 2.

The algorithm requires $O(kn^3)$ operations if all edge lengths of the input graph are nonnegative, and $O(kn^4)$ operations if some edge lengths are negative.

The subroutine KSHOT2 below will calculate the k shortest path lengths, and subroutine GETPTH will output the k shortest paths.

C. Subroutine KSHOT2 parameters

Input:

N	Number of nodes.
M	Number of edges.
INODE, JNODE	Each is an integer vector of length M; the ith edge is directed from INODE(i) to JNODE(i).
ARCLEN	Integer vector of length M; ARCLEN(i) is the length of the ith edge.
KPATHS	The required number of shortest paths, KPATHS > 0.
KP3	Equal to KPATHS + 3.
MAXQUE	Size of the auxiliary storage; in most cases, it is sufficient to set MAXQUE equal to two times KPATHS; see parameter IFLAG below.
ISORCE, ISINK	It is required to find the KPATHS shortest paths from node ISORCE to node ISINK.

Output:

IFLAG	IFLAG = 0	for normal termination;
	IFLAG = 1	if no paths are found;
	IFLAG = 2	if the size of the auxiliary arrays CROSAR and NEXTRD are not large enough, in this case, increase their sizes by using a larger value of MAXQUE and rerun the program.
IPATHS	The actual number of shortest paths found by KSHOT2, IPATHS ≤ KPATHS.	
PTHLEN	Integer vector of length KP3; the shortest path lengths are stored in PTHLEN(2), PTHLEN(3), . . . , PTHLEN(IPATHS + 1).	

Working storages:

ARCDIR	Integer vector of length N; array of shortest path tree.
TRDIST	Integer vector of length N; array of node information.
ARCNOD	Integer vector of length N; array of node information.
ARCFWD	Integer vector of length N; first edge in forward star.
ARCBWD	Integer vector of length N; first edge in backward star.
AUXSTG	Integer vector of length N; auxiliary storage.
AUXDIS	Integer vector of length N; auxiliary node array.
AUXTRE	Integer vector of length N; auxiliary node array.

AUXLNK	Integer vector of length N; auxiliary node array.
NXTFWD	Integer vector of length M; links of node chain.
NXTBWD	Integer vector of length M; links of node chain.
QUFIRP	Integer vector of length KP3; description of the first path in a queue.
QUNXTP	Integer vector of length KP3; next entry in a queue.
NQTAB	Integer vector of length KP3; search array.
CROSAR	Integer vector of length MAXQUE; cross edge of the path.
NEXTRD	Integer vector of length MAXQUE: next entry of the pool storage.

Subroutine GETPTH parameters

Input:

N	Number of nodes.
M	Number of edges.
K	Generate the Kth shortest path, $K \leq$ IPATHS, where IPATHS is an output from KSHOT2.
KP3	Equal to KPATHS + 3; see parameters in KSHOT2.
MAXQUE	See parameters in KSHOT2.
INODE, JNODE, PTHLEN, ARCDIR, QUFIRP, QUNXTP, CROSAR, NEXTRD	Output from subroutine KSHOT2.

Output:

NUMP	Number of edges in the path.
LENGTH	The length of the path.
ARCNOD	Integer vector of length N; the edges of the Kth shortest path are stored in ARCNOD(1), ARCNOD(2), . . . , ARCNOD(NUMP).

D. Test example

Find the four shortest paths from node 5 to node 3 in the following graph.

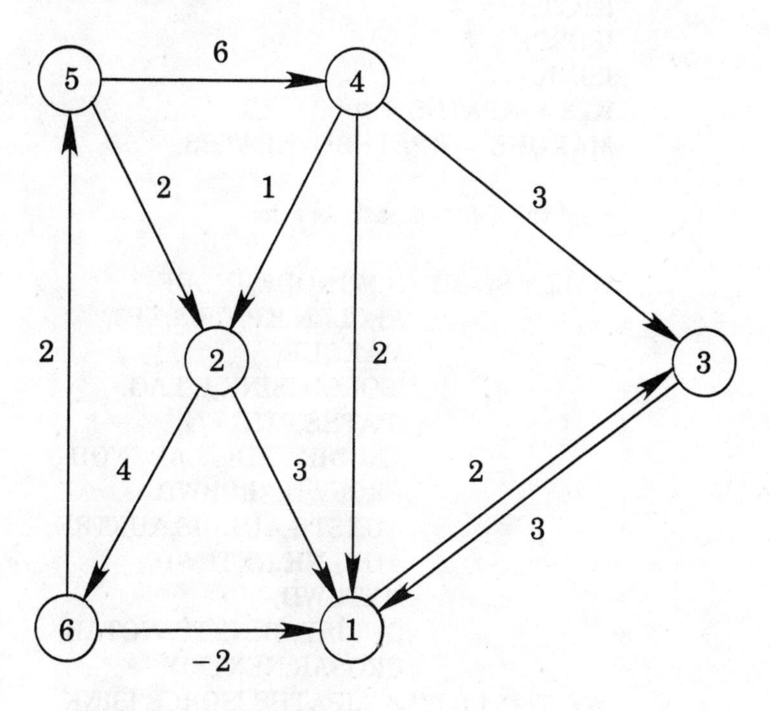

E. Program listing

MAIN PROGRAM

```
      INTEGER INODE(11),JNODE(11),
     +        ARCLEN(11),PTHLEN(7),
```

```
     +              ARCDIR(6),TRDIST(6),ARCNOD(6),
     +              ARCFWD(6),ARCBWD(6),
     +              AUXSTG(6),AUXDIS(6),
     +              AUXTRE(6),AUKLNK(6),
     +              NXTFWD(11),NXTBWD(11),
     +              QUFIRP(7),QUNXTP(7),NQTAB(7),
     +              CROSAR(8),NEXTRD(8)
      DATA INODE /4, 3, 6, 4, 2, 5, 6, 5, 1, 2, 4/,
     +       JNODE /2, 1, 5, 3, 1, 4, 1, 2, 3, 6, 1/,
     +       ARCLEN /1, 3, 2, 3, 3, 6, -2, 2, 2, 4, 2/
C
      N = 6
      M = 11
      KPATHS = 4
      ISORCE = 5
      ISINK = 3
      KP3 = KPATHS + 3
      MAXQUE = KPATHS + KPATHS
C
C     Find the shortest path lengths
C
      CALL KSHOT2 (N,M,INODE,JNODE,
     +                   ARCLEN,KPATHS,KP3,
     +                   MAXQUE,
     +                   ISORCE,ISINK,IFLAG,
     +                   IPATHS,PTHLEN,
     +                   ARCDIR,TRDIST,ARCNOD,
     +                   ARCFWD,ARCBWD,
     +                   AUXSTG,AUXDIS,AUXTRE,
     +                   AUXLNK,NXTFWD,
     +                   NXTBWD,
     +                   QUFIRP,QUNXTP, NQTAB,
     +                   CROSAR,NEXTRD)
      WRITE(*,10) IFLAG,IPATHS,ISORCE,ISINK
   10 FORMAT(/' IFLAG =',I3//
     +          ' THE FIRST',I3,' SHORTEST',
     +          ' PATHS',
```

```
     +              ' FROM NODE',I3,' TO NODE',I3,
     +              ' WERE FOUND'//
     +              ' THERE PATH LENGTHS ARE:'/)
      J = IPATHS + 1
      DO 20 I = 2, J
 20     WRITE(*,30) PTHLEN(I)
 30     FORMAT(8X,I6)
C
C     Output the paths
C
      DO 50 K = 1, IPATHS
        CALL GETPTH (N,M,K,KP3,MAXQUE,
     +                    INODE,JNODE,PTHLEN,
     +                    ARCDIR,QUFIRP,
     +                    QUNXTP,CROSAR,
     +                    NEXTRD,
     +                    NUMP,LENGTH,ARCNOD)
        WRITE(*,40) K,NUMP,LENGTH,
     +    (ARCNOD(J),J=1,NUMP)
 40     FORMAT(//' SHORTEST PATH',
     +              ' NUMBER',I3,':'//
     +              ' THE NUMBER OF EDGES IN',
     +              ' THE PATH =',I4/
     +              ' THE LENGTH OF THE',
     +              ' PATH = ',I4/
     +              ' THE EDGES IN THE',
     +              ' PATH:',20I4/)
 50   CONTINUE
C
      STOP
      END
```

OUTPUT RESULTS

```
IFLAG = 0
THE FIRST 4 SHORTEST PATHS FROM NODE 5
  TO NODE 3 WERE FOUND
```

```
THEIR PATH LENGTHS ARE:
                6
                7
                9
               10
SHORTEST PATH NUMBER 1:
 THE NUMBER OF EDGES IN THE PATH = 4
          THE LENGTH OF THE PATH = 6
            THE EDGES IN THE PATH:   8 10 7 9
SHORTEST PATH NUMBER 2:
 THE NUMBER OF EDGES IN THE PATH = 3
          THE LENGTH OF THE PATH = 7
            THE EDGES IN THE PATH:   8 5 9
SHORTEST PATH NUMBER 3:
 THE NUMBER OF EDGES IN THE PATH = 2
          THE LENGTH OF THE PATH = 9
            THE EDGES IN THE PATH:   6 4
SHORTEST PATH NUMBER 4:
 THE NUMBER OF EDGES IN THE PATH =  3
          THE LENGTH OF THE PATH = 10
            THE EDGES IN THE PATH:    6 11 9
```

```
      SUBROUTINE KSHOT2 (N,M,INODE,JNODE,
     +                   ARCLEN,KPATHS,
     +                   KP3,MAXQUE,
     +                   ISORCE,ISINK,
     +                   IFLAG,IPATHS,
     +                   PTHLEN,
     +                   ARCDIR,TRDIST,
     +                   ARCNOD,ARCFWD
     +                   ARCBWD,
     +                   AUXSTG,AUXDIS,
     +                   AUXTRE,AUXLNK,
     +                   NXTFWD,NXTBWD,
     +                   QUFIRP,QUNXTP,
     +                   NQTAB,CROSAR,
```

```
     +                              NEXTRD)
C
C      Find the KPATHS shortest distinct path lengths
C        from ISORCE to ISINK without repeated
C        nodes
C
       INTEGER INODE(M),JNODE(M),
     +           ARCLEN(M),PTHLEN(KP3),
     +           ARCDIR(N),TRDIST(N),
     +           ARCNOD(N),ARCFWD(N),
     +           ARCBWD(N),
     +           AUXSTG(N),AUXDIS(N),
     +           AUXTRE(N),AUXLNK(N),
     +           NXTFWD(M),NXTBWD(M),
     +           QUFIRP(KP3),QUNXTP(KP3),
     +           NQTAB(KP3),CROSAR(MAXQUE),
     +           NEXTRD(MAXQUE)
       LOGICAL FORWRD,LASTNO,NOROOM,
     +           FINEM1,FINEM2,INITSP,GOON,
     +           LOOPON,LASTA,LASTB,NOSTG,
     +           RDFULL
C
C      Set up the network representation
C
       IFLAG = 0
       DO 10 I = 1, N
         ARCFWD(I) = 0
         ARCBWD(I) = 0
  10   CONTINUE
C
       LARGE = 1
       DO 20 I = 1, M
         LARGE = LARGE + ARCLEN(I)
         ITAIL = INODE(I)
         IHEAD = JNODE(I)
         NXTFWD(I) = ARCFWD(ITAIL)
         ARCFWD(ITAIL) = I
         NXTBWD(I) = ARCBWD(IHEAD)
         ARCBWD(IHEAD) = I
```

```
   20    CONTINUE
C
C        Initialize
C
         DO 30 I = 1, N
   20      AUXDIS(I) = LARGE
C
         DO 40 I = 1, KP3
   40      QUFIRP(I) = 0
C
         DO 50 I = 1, MAXQUE
   50      NEXTRD(I) = I
         NEXTRD(MAXQUE) = 0
C
C        Build the shortest distance tree
C
C        TRDIST(i) is used to store the shortest distance
C          of node i from ISORCE; ARCDIR(i) will
C          contain the tree edge coming to node i; it is
C          negative if the direction of the edge is towards
C          ISORCE, and it is zero if not reachable.
C
         DO 60 I = 1, N
           TRDIST(I) = LARGE
           ARCDIR(I) = 0
           ARCNOD(I) = 0
   60    CONTINUE
         TRDIST(ISORCE) = 0
         ARCDIR(ISORCE) = ISORCE
         ARCNOD(ISORCE) = ISORCE
         J = ISORCE
         NODE1 = ISORCE
C
C        Examine neighbors of node J
C
   70    IEDGE1 = ARCFWD(J)
         FORWRD = .TRUE.
         LASTNO = .FALSE.
C
```

```
80    IF (IEDGE1 .EQ. 0) THEN
        LASTNO = .TRUE.
      ELSE
        LENGTH = TRDIST(J) + ARCLEN(IEDGE1)
        IF (FORWRD) THEN
          NODE2 = JNODE(IEDGE1)
          IEDGE2 = IEDGE1
        ELSE
          NODE2 = INODE(IEDGE1)
          IEDGE2 = -IEDGE1
        ENDIF
        IF (LENGTH .LT. TRDIST(NODE2)) THEN
          TRDIST(NODE2) = LENGTH
          ARCDIR(NODE2) = IEDGE2
          IF (ARCNOD(NODE2) .EQ. 0) THEN
            ARCNOD(NODE1) = NODE2
            ARCNOD(NODE2) = NODE2
            NODE1 = NODE2
          ELSE
            IF (ARCNOD(NODE2) .LT. 0) THEN
              ARCNOD(NODE2) = ARCNOD(J)
              ARCNOD(J) = NODE2
              IF (NODE1 .EQ. J) THEN
                NODE1 = NODE2
                ARCNOD(NODE2) = NODE2
              ENDIF
            ENDIF
          ENDIF
        ENDIF
        IF (FORWRD) THEN
          IEDGE1 = NXTFWD(IEDGE1)
        ELSE
          IEDGE1 = NXTBWD(IEDGE1)
        ENDIF
      ENDIF
C
      IF (.NOT. LASTNO) GOTO 80
C
      JJ = J
```

```
          J = ARCNOD(J)
          ARCNOD(JJ) = -1
          IF (J .NE. JJ) GOTO 70
C
C         Finish building the shortest distance tree
C
          IPATHS = 0
          NOROOM = .FALSE.
          IF (ARCDIR(ISINK) .EQ. 0) THEN
            IFLAG = 1
            RETURN
          ENDIF
C
C         Initialize the storage pool
C
          I = 1
   90       QUNXTP(I) = I
            I = I + 1
          IF (I .LE. KPATHS + 2) GOTO 90
          QUNXTP(KPATHS + 3) = 0
C
C         Initialize the priority queue
C
          LENTAB = KP3
          LOW = -LARGE
          IHIGH = LARGE
          NQOP1 = LENTAB
          NQOP2 = 0
          NQSIZE = 0
          NQIN = 0
          NQOUT1 = 0
          NQOUT2 = 0
C
C         Obtain an entry from storage pool
C
          INDEX1 = QUNXTP(1)
          QUNXTP(1) = QUNXTP(INDEX1 + 1)
          INDEX2 = QUNXTP(1)
```

```
      QUNXTP(1) = QUNXTP(INDEX2 + 1)
      PTHLEN(INDEX1 + 1) = LOW
      QUNXTP(INDEX1 + 1) = INDEX2
      PTHLEN(INDEX2 + 1) = IHIGH
      QUNXTP(INDEX2 + 1) = 0
      NQP1 = 0
      NQP2 = 1
      NQTAB(1) = INDEX1
      NQTAB(2) = INDEX2
      NQFIRS = IHIGH
      NQLAST = LOW
C
C     Set the shortest path to the queue
C
      IPOOL1 = QUNXTP(1)
      QUNXTP(1) = QUNXTP(IPOOL1 + 1)
      IPOOL2 = IPOOL1
      INCRS1 = NEXTRD(1)
      NEXTRD(1) = NEXTRD(INCRS1 + 1)
      CROSAR(INCRS1 + 1) = ARCDIR(ISINK)
      NEXTRD(INCRS1 + 1) = 0
      PTHLEN(IPOOL1 + 1) = TRDIST(ISINK)
      QUFIRP(IPOOL1 + 1) = INCRS1
      IPARM = IPOOL1
      ICALL = 0
C
C     Insert IPARM into the priority queue
C
 100  IORDER = PTHLEN(IPARM + 1)
      IOP1 = NQP1
      IOP2 = NQP2
C
 110  IF (IOP2 - IOP1 .GT. 1) THEN
         IOP3 = (IOP1 + IOP2) / 2
         IF (IORDER .GT. PTHLEN(NQTAB
     +      (IOP3 + 1) + 1)) THEN
            IOP1 = IOP3
         ELSE
```

```
            IOP2 = IOP3
          ENDIF
          GOTO 110
        ENDIF
C
C       Linear search starting from NQTAB(IOP1 + 1)
C
        INDEX1 = NQTAB(IOP1 + 1)
 120      INDEX2 = INDEX1
          INDEX1 = QUNXTP(INDEX1 + 1)
        IF (PTHLEN(INDEX1 + 1) .LE. IORDER)
     +    GOTO 120
C
C       Insert between INDEX1 and INDEX2
C
        QUNXTP(INDEX2 + 1) = IPARM
        QUNXTP(IPARM + 1) = INDEX1
C
C       Update data in the queue
C
        NQSIZE = NQSIZE + 1
        NQIN = NQIN + 1
        NQOP1 = NQOP1 - 1
C
        IF (NQSIZE .EQ. 1) THEN
          NQFIRS = IORDER
          NQLAST = IORDER
        ELSE
          IF (IORDER .GT. NQLAST) THEN
            NQLAST = IORDER
          ELSE
            IF (IORDER .LT. NQFIRS)
     +        NQFIRS = IORDER
          ENDIF
        ENDIF
C
        IF (NQOP1 .LE. 0) THEN
C
```

```
C      Reorganize
C
       INDEX1 = NQTAB(NQP1 + 1)
       NQTAB(1) = INDEX1
       NQP1 = 0
       INDEX2 = NQTAB(NQP2 + 1)
       J3 = NQSIZE / LENTAB
       J2 = J3 + 1
       J1 = MOD(NQSIZE,LENTAB)
       IF (J1 .GT. 0) THEN
         DO 140 IOP2 = 1, J1
           DO 130 I = 1, J2
130          INDEX1 = QUNXTP(INDEX1 + 1)
           NQTAB(IOP2 + 1) = INDEX1
140      CONTINUE
       ENDIF
       IF (J3 .GT. 0) THEN
         IOP2 = J1 + 1
150      IF (IOP2 .LE. LENTAB - 1) THEN
           DO 160 I = 1, J3
160          INDEX1 = QUNXTP(INDEX1 + 1)
           NQTAB(IOP2 + 1) = INDEX1
           IOP2 = IOP2 + 1
           GOTO 150
         ENDIF
       ENDIF
       NQP2 = IOP2
       NQTAB(NQP2 + 1) = INDEX2
       NQOP2 = NQOP2 + 1
       NQOP1 = NQSIZE / 2
       IF (NQOP1 .LT. LENTAB)
     +   NQOP1 = LENTAB
     ENDIF
     IF (ICALL .GT. 0) GOTO 430
C
     ILEN1 = 0
     MARK1 = 0
     INITSP = .TRUE.
```

```
      DO 170 I = 1, N
170     ARCNOD(I) = 0
C
C     Process the next path
C
180   MARK1 = MARK1 + 2
      MARK2 = MARK1
      MSHADE = MARK1 + 1
C
C     Obtain the first entry from the priority queue
C
      IF (NQSIZE .GT. 0) THEN
        INDEX2 = NQTAB(NQP1 + 1)
        INDEX1 = QUNXTP(INDEX2 + 1)
        QUNXTP(INDEX2 + 1)
     +     = QUNXTP(INDEX1 + 1)
        NQFIRS = PTHLEN
     +     (QUNXTP(INDEX1 + 1) + 1)
        IF (INDEX1 .EQ. NQTAB(NQP1 + 2)) THEN
          NQP1 = NQP1 + 1
          NQTAB(NQP1 + 1) = INDEX2
        ENDIF
        NQOP1 = NQOP1 - 1
        NQSIZE = NQSIZE - 1
        NQOUT1 = NQOUT1 + 1
        IPOOL3 = INDEX1
      ELSE
        IPOOL3 = 0
      ENDIF
      IF (IPOOL3 .EQ. 0) THEN
C
C       No more paths in queue; stop.
C
        NOROOM = NOROOM .AND.
     +     (IPATHS .LT. KPATHS)
        GOTO 450
      ENDIF
      QUNXTP(IPOOL2 + 1) = IPOOL3
      IPOOL2 = IPOOL3
```

```
      IPATHS = IPATHS + 1
      IF (IPATHS .GT. KPATHS) THEN
        NOROOM = .FALSE.
        IPATHS = IPATHS - 1
        GOTO 450
      ENDIF
      ILEN2 = PTHLEN(IPOOL3 + 1)
      IQUFIR = QUFIRP(IPOOL3 + 1)
      IF (ILEN2 .LT. ILEN1) GOTO 450
      ILEN1 = ILEN2
C
C     Examine the tail of the edge
C
      INCRS2 = IQUFIR
      NCRS2 = ISORCE
      NODEP1 = N + 1
C
C     Obtain data of next path
C
 190  JUMP = 1
      GOTO 500
C
 200  ILEN2 = ILEN2 - ARCLEN(INCRS4)
      NODEP1 = NODEP1 - 1
      AUXSTG(NODEP1) = INCRS1
 210  IF (NCRS1 .NE. NCRS3) THEN
        J = IABS(ARCDIR(NCRS1))
        ILEN2 = ILEN2 - ARCLEN(J)
        IF (ARCDIR(NCRS1) .GT. 0) THEN
          NCRS1 = INODE(J)
        ELSE
          NCRS1 = JNODE(J)
        ENDIF
        GOTO 210
      ENDIF
      IF (.NOT. FINEM1) GOTO 190
C
C     Store the tail of the edge
C
```

```
         NODEP2 = NODEP1
         FINEM2 = FINEM1
C
C        Obtain data of next path
C
         JUMP = 2
         GOTO 500
C
 220     IF (LINKST .EQ. 2) THEN
            NODEP2 = NODEP2 - 1
            AUXSTG(NODEP2) = INCRS4
            ARCLEN(INCRS4) = ARCLEN(INCRS4)
     +         + LARGE
            FINEM2 = FINEM1
C
C           Obtain data of next path
C
            JUMP = 2
            GOTO 500
         ENDIF
C
C        Close the edge on the shortest path
C
         FINEM2 = FINEM2 .AND. (LINKST .NE. 3)
         IF (FINEM2) THEN
            IEDGE = IABS(ARCDIR(NCRS3))
            NODEP2 = NODEP2 - 1
            AUXSTG(NODEP2) = IEDGE
            ARCLEN(IEDGE) = ARCLEN(IEDGE)
     +         + LARGE
         ENDIF
C
C        Mark more nodes
C
 230     IF (LINKST .NE. 3) THEN
C
 240        ARCNOD(NCRS2) = MARK2
 250          IF (NCRS1 .NE. NCRS3) THEN
                 ARCNOD(NCRS1) = MARK2
```

```
          IF (ARCDIR(NCRS1) .GT. 0) THEN
            NCRS1 = INODE(ARCDIR(NCRS1))
          ELSE
            NCRS1 = JNODE(-ARCDIR(NCRS1))
          ENDIF
          GOTO 250
        ENDIF
        JUMP = 3
        GOTO 500
C
 260    IF (LINKST .EQ. 1) GOTO 240
C
 270    IF (LINKST .EQ. 2) THEN
          JUMP = 4
          GOTO 500
        ENDIF
        GOTO 230
      ENDIF
C
C     Generate descendants of the tail of the edge
C
      NODEP3 = NODEP1
      INCRS1 = AUXSTG(NODEP3)
      JND1 = CROSAR(INCRS1 + 1)
C
C     Obtain the first node of the edge traversing
C       forward
C
      IF (JND1 .LT. 0) THEN
        JND2 = INODE(-JND1)
      ELSE
        JND2 = JNODE(JND1)
      ENDIF
C
C     Process a section
C
 280  NODEP3 = NODEP3 + 1
      JTERM = JND2
      JEDGE = JND1
```

```
      IF (NODEP3 .GT. N) THEN
        JND2 = ISORCE
      ELSE
        INCRS2 = AUXSTG(NODEP3)
        JND1 = CROSAR(INCRS2 + 1)
        IF (-JND1 .GT. 0) THEN
          JND2 = INODE(-JND1)
        ELSE
          JND2 = JNODE(JND1)
        ENDIF
      ENDIF
C
C     Process a node
C
290   MARK1 = MARK 1 + 2
      ITREAR = MARK1
      IEXAM = MARK1 + 1
      IEDGE = IABS(JEDGE)
      ARCLEN(IEDGE) = ARCLEN(IEDGE)
     +  + LARGE
      IF (INITSP) INITSP = NQIN .LT. KPATHS
      IF (INITSP) THEN
        IUPBND = LARGE
      ELSE
        IUPBND = NQLAST
      ENDIF
C
C     Obtain the restricted shortest path from
C        ISORCE to JTERM
C
      LENRST = IUPBND
      IBEDGE = 0
      AUXDIS(JTERM) = 0
      AUXTRE(JTERM) = 0
      AUXLNK(JTERM) = 0
      JEM1 = JTERM
      JEM2 = JEM1
C
```

```
C       Examine next node
C
 300    NJEM1 = JEM1
        IAUXD1 = AUXDIS(NJEM1)
        JEM1 = AUXLNK(NJEM1)
        ARCNOD(NJEM1) = ITREAR
        IF (IAUXD1 + TRDIST(NJEM1)
     +  + ILEN2 .GE. LENRST) GOTO 360
        GOON = .TRUE.
        LASTA = .FALSE.
        IEDGE1 = ARCBWD(NJEM1)
C
C       Loop through edges from NJEM1
C
 310    IF (IEDGE1 .EQ. 0) THEN
          LASTA = .TRUE.
        ELSE
C
C         Process the edge IEDGE1
C
          IAUXD2 = IAUXD1 + ARCLEN(IEDGE1)
          IF (GOON) THEN
            NJEM2 = INODE(IEDGE1)
            IEDGE2 = IEDGE1
            IEDGE1 = NXTBWD(IEDGE1)
          ELSE
            NJEM2 = JNODE(IEDGE1)
            IEDGE2 = -IEDGE1
            IEDGE1 = NXTFWD(IEDGE1)
          ENDIF
          IF (ARCNOD(NJEM2) .NE. MARK2) THEN
            IAUXD3 = IAUXD2 + ILEN2
     +        + TRDIST(NJEM2)
            IF (IAUXD3 .GE. LENRST) GOTO 350
            IF (ARCNOD(NJEM2) .LT. MARK2) THEN
              IF (ARCDIR(NJEM2) + IEDGE2 .EQ. 0)
     +          THEN
                ARCNOD(NJEM2) = MSHADE
                GOTO340
```

```
              ENDIF
C
C             Examine the status of the path
C
              LOOPON = .TRUE.
              NJEM3 = NJEM2
 320          IF (LOOPON .AND.
     +          (NJEM3 .NE. ISORCE)) THEN
                IF (ARCNOD(NJEM3) .LT. MARK2)
     +            THEN
                  J = ARCDIR(NJEM3)
                  IF (J .GT. 0) THEN
                    NJEM3 = INODE(J)
                  ELSE
                    NJEM3 = JNODE(-J)
                  ENDIF
                ELSE
                  LOOPON = .FALSE.
                ENDIF
                GOTO 320
              ENDIF
              IF (LOOPON) THEN
C
C               Better path found
C
                LENRST = IAUXD3
                IBEDGE = IEDGE2
                GOTO 350
              ELSE
                NJEM3 = NJEM2
                LASTB = .FALSE.
 330            IF (ARCNOD(NJEM3) .LT. MARK2)
     +            THEN
                  ARCNOD(NJEM3) = MSHADE
                  J = ARCDIR(NJEM3)
                  IF (J .GT. 0) THEN
                    NJEM3 = INODE(J)
                  ELSE
```

```
                NJEM3 = JNODE(-J)
              ENDIF
            ELSE
              LASTB = .TRUE.
            ENDIF
            IF (.NOT. LASTB) GOTO 330
          ENDIF
        ENDIF
340     IF ((ARCNOD(NJEM2) .LT. ITREAR) .OR.
     +      (IAUXD2 .LT. AUXDIS(NJEM2)))
     +        THEN
C
C         Update node NJEM2
C
          AUXDIS(NJEM2) = IAUXD2
          AUXTRE(NJEM2) = IEDGE2
          IF (ARCNOD(NJEM2) .NE. IEXAM)
     +      THEN
            ARCNOD(NJEM2) = IEXAM
            IF (JEM1 .EQ. 0) THEN
              JEM1 = NJEM2
              JEM2 = NJEM2
              AUXLNK(NJEM2) = 0
            ELSE
              IF (ARCNOD(NJEM2) .EQ..ITREAR)
     +          THEN
                AUXLNK(NJEM2) = JEM1
                JEM1 = NJEM2
              ELSE
                AUXLNK(NJEM2) = 0
                AUXLNK(JEM2) = NJEM2
                JEM2 = NJEM2
              ENDIF
            ENDIF
          ENDIF
        ENDIF
      ENDIF
    ENDIF
```

```
350    IF (.NOT. LASTA) GOTO 310
360    IF (JEM1 .GT. 0) GOTO 300
       ARCNOD(JTERM) = MARK2
C
C      Finish processing the restricted path
C
       IF ((IBEDGE .NE. 0)
     +   .AND. (LENRST .LT. IUPBND)) THEN
         IDET = 0
         ICEDGE = IBEDGE
370        IF (ICEDGE .GT. 0) THEN
             IDEDGE = JNODE(ICEDGE)
           ELSE
             IDEDGE = INODE(-ICEDGE)
           ENDIF
           IF ((ICEDGE .NE. ARCDIR(IDEDGE)) .OR.
     +         (IDEDGE .EQ. JTERM)) THEN
             IDET = IDET + 1
             AUXSTG(IDET) = ICEDGE
           ENDIF
           ICEDGE = AUXTRE(IDEDGE)
         IF (ICEDGE .NE. 0) GOTO 370
C
C      Restore the path data
C      NOSTG will be TRUE if the arrays CROSAR
C        and NEXTRD need more space
C
         IDET1 = IDET
         INRD1 = NEXTRD(1)
         INQLAS = LARGE
         NOSTG = .FALSE.
380      IF ((IDET1 .GT. 0) .AND.
     +       (INRD1 .GT. 0)) THEN
           IDET1 = IDET1 - 1
           INRD1 = NEXTRD(INRD1 + 1)
           GOTO 380
         ENDIF
         RDFULL = (.NOT. INITSP)
     +       .AND. (IPATHS + NQSIZE .GE. KPATHS)
```

```
390       IF (RDFULL .OR. (IDET1 .GT. 0)) THEN
C
C           Remove the last path from the queue
C
            INQLAS = NQLAST
            NOROOM = .TRUE.
            RDFULL = .FALSE.
C
C           Get the last entry from the priority queue
C
            IF (NQSIZE .GT. 0) THEN
              INDEX4 = NQTAB(NQP2 + 1)
              INDEX3 = NQTAB(NQP2)
              IF (QUNXTP(INDEX3 + 1)
     +          .EQ. INDEX4) THEN
                NQP2 = NQP2 - 1
                NQTAB(NQP2 + 1) = INDEX4
                INDEX3 = NQTAB(NQP2)
              ENDIF
              INDEX2 = INDEX3
400           IF (INDEX3 .NE. INDEX4) THEN
                INDEX1 = INDEX2
                INDEX2 = INDEX3
                INDEX3 = QUNXTP(INDEX3 + 1)
                GOTO 400
              ENDIF
              QUNXTP(INDEX1 + 1) = INDEX4
              NQLAST = PTHLEN(INDEX1 + 1)
              NQOP1 = NQOP1 - 1
              NRDSIZ = INDEX2
              NQSIZE = NQSIZE - 1
              NQOUT2 = NQOUT2 + 1
            ELSE
              NRDSIZ = 0
            ENDIF
C
            IF (NRDSIZ .EQ. 0) THEN
              NOSTG = .TRUE.
              GOTO 430
```

```
          ENDIF
          INRD1 = QUFIRP(NRDSIZ + 1)
  410     IF (INRD1 .GT. 0) THEN
            J = INRD1 + 1
            IDET1 = IDET1 - 1
            INRD2 = INRD1
            INRD1 = NEXTRD(J)
            NEXTRD(J) = NEXTRD(1)
            NEXTRD(1) = INRD2
            GOTO 410
          ENDIF
C
C         Put the entry NRDSIZ into storage pool
C
          QUNXTP(NRDSIZ + 1) = QUNXTP(1)
          QUNXTP(1) = NRDSIZ
          GOTO 390
        ENDIF
C
C       Build the entries of CROSAR and NEXTRD
C
        IF (LENRST .GE. INQLAS) GOTO 430
        INRD2 = -INCRS1
        IDET1 = IDET
  420   IF (IDET1 .GT. 0) THEN
          INRD1 = NEXTRD(1)
          NEXTRD(1) = NEXTRD(INRD1 + 1)
          CROSAR(INRD1 + 1) = AUXSTG(IDET1)
          NEXTRD(INRD1 + 1) = INRD2
          INRD2 = INRD1
          IDET1 = IDET1 - 1
          GOTO 420
        ENDIF
C
C       Obtain the entry NRDSIZ from storage pool
C
        NRDSIZ = QUNXTP(1)
```

```
         QUNXTP(1) = QUNXTP(NRDSIZ + 1)
         PTHLEN(NRDSIZ + 1) = LENRST
         QUFIRP(NRDSIZ + 1) = INRD2
         IPARM = NRDSIZ
         ICALL = 1
         GOTO 100
C
 430     IF (NOSTG) THEN
           IFLAG = 2
           GOTO 450
         ENDIF
       ENDIF
C
       ARCLEN(IEDGE) = ARCLEN(IEDGE)
     +   - LARGE
       ILEN2 = ILEN2 + ARCLEN(IEDGE)
       IF (JTERM .NE. JND2) THEN
         IF (JEDGE .GT. 0) THEN
           JTERM = INODE(JEDGE)
         ELSE
           JTERM = JNODE(-JEDGE)
         ENDIF
         JEDGE = ARCDIR(JTERM)
       ENDIF
       IF (JTERM .NE. JND2) GOTO 290
C
       INCRS1 = INCRS2
       IF (NODEP3 .LE. N) GOTO 280
C
C      Restore the join edges
C
 440   IF (NODEP2 .LE. NODEP1 - 1) THEN
         J = AUXSTG(NODEP2)
         ARCLEN(J) = ARCLEN(J) - LARGE
         NODEP2 = NODEP2 + 1
         GOTO 440
       ENDIF
```

```
C
C       Repeat with the next path
C
        GOTO 180
C
C       SORT THE PATHS
C
 450    IHD2 = IPOOL1
        ICNT = 0
 460      IHD1 = IHD2
          ICNT = ICNT + 1
          IHD2 = QUNXTP(IHD1 + 1)
          QUNXTP(IHD1 + 1) = ICNT
        IF (IHD1 .NE. IPOOL2) GOTO 460
C
C       Release all queue entries to the storage pool
C
        J = NQTAB(NQP2 + 1)
        QUNXTP(J + 1) = QUNXTP(1)
        QUNXTP(1) = NQTAB(NQP1 + 1)
        NQP1 = 0
        NQP2 = 0
C
        IHD2 = 0
 470      J = IHD2 + 1
          IHD2 = QUNXTP(J)
          QUNXTP(J) = 0
        IF (IHD2 .NE. 0) GOTO 470
C
C       Exchanging records
C
        JJ = KPATHS + 2
        DO 490 I = 1, JJ
 480      IF ((QUNXTP(I + 1) .GT. 0) .AND.
     +        (QUNXTP(I + 1) .NE. I)) THEN
            TQUORD = PTHLEN(I + 1)
            TQUFIR = QUFIRP(I + 1)
            TQUNXT = QUNXTP(I + 1)
            J = QUNXTP(I + 1) + 1
```

```
            PTHLEN(I + 1) = PTHLEN(J)
            QUFIRP(I + 1) = QUFIRP(J)
            QUNXTP(I + 1) = QUNXTP(J)
            PTHLEN(TQUNXT + 1) = TQUORD
            QUFIRP(TQUNXT + 1) = TQUFIR
            QUNXTP(TQUNXT + 1) = TQUNXT
            GOTO 480
          ENDIF
  490   CONTINUE
        PTHLEN(1) = ISORCE
        QUFIRP(1) = ISINK
        QUNXTP(1) = IPATHS
        RETURN
C
C       Obtain data for the next path
C
  500   IF (INCRS2 .EQ. 0) THEN
          LINKST = 3
        ELSE
          NCRS3 = NCRS2
          INCRS1 = INCRS2
          J = IABS(INCRS1) + 1
          INCRS3 = CROSAR(J)
          INCRS2 = NEXTRD(J)
          IF (INCRS3 .GT. 0) THEN
            NCRS1 = INODE(INCRS3)
            NCRS2 = JNODE(INCRS3)
            INCRS4 = INCRS3
          ELSE
            NCRS1 = JNODE(-INCRS3)
            NCRS2 = INODE(-INCRS3)
            INCRS4 = -INCRS3
          ENDIF
          FINEM1 = INCRS2 .LE. 0
          IF (NCRS2 .EQ. NCRS3) THEN
            LINKST = 2
          ELSE
            LINKST = 1
          ENDIF
```

```
      ENDIF
      GOTO (200,220,260,270), JUMP
C
      RETURN
      END


      SUBROUTINE GETPTH (N,M,K,KP3,
     +                            MAXQUE,INODE,
     +                            JNODE,PTHLEN,
     +                            ARCDIR,QUFIRP,
     +                            QUNXTP,CROSAR,
     +                            NEXTRD, NUMP,
     +                            LENGTH, ARCNOD)
C
C     Retrieve the edges of the Kth shortest path (this
C       subprogram is used after subroutine KSHOT2)
C
      INTEGER INODE(M),JNODE(M),
     +            PTHLEN(KP3),ARCDIR(N),
     +            QUFIRP(KP3),
     +            QUNXTP(KP3),
     +            CROSAR(MAXQUE),
     +            NEXTRD(MAXQUE),ARCNOD(N)
C
      NUMBER = 0
      IF ((K .LE. 0) .OR. (K .GT. QUNXTP(1))) THEN
        NUMP = NUMBER
        RETURN
      ENDIF
      INDEX2 = PTHLEN(1)
      LENGTH = PTHLEN(K + 1)
      ISUB3 = QUFIRP(K + 1)
   10 IF (ISUB3 .NE. 0) THEN
        JSUB = IABS(ISUB3) + 1
        INDEX3 = INDEX2
        IF (CROSAR(JSUB) .GT. 0) THEN
          INDEX1 = INODE(CROSAR(JSUB))
          INDEX2 = JNODE(CROSAR(JSUB))
```

```
          ELSE
            INDEX1 = JNODE(-CROSAR(JSUB))
            INDEX2 = INODE(-CROSAR(JSUB))
          ENDIF
          IF (INDEX2 .NE. INDEX3) THEN
C
C           Store the edges
C
            ISUB2 = N
            ARCNOD(ISUB2) = CROSAR(JSUB)
   20       IF (INDEX1 .NE. INDEX3) THEN
              ISUB1 = ARCDIR(INDEX1)
              ISUB2 = ISUB2 - 1
              IF (ISUB2 .GT. 0) THEN
                ARCNOD(ISUB2) = ISUB1
              ELSE
                NUMP = ISUB1
              ENDIF
              IF (ISUB1 .GT. 0) THEN
                INDEX1 = INODE(ISUB1)
              ELSE
                INDEX1 = JNODE(-ISUB1)
              ENDIF
              GOTO 20
            ENDIF
   30       IF (ISUB2 .LE. N) THEN
              NUMBER = NUMBER + 1
              ARCNOD(NUMBER = ARCNOD(ISUB2)
              ISUB2 = ISUB2 + 1
              GOTO 30
            ENDIF
          ENDIF
          ISUB3 = NEXTRD(JSUB)
          GOTO 10
        ENDIF
C
        NUMP = NUMBER
        RETURN
        END
```

3

TRAVERSABILITY

3-1 Euler Circuit

A. Problem description

An *Euler circuit* of an undirected graph is a walk which starts and ends at the same node and uses each edge exactly once. In the case of a directed graph, the walk is directed. A graph is *Eulerian* if it has an Euler circuit. The problem is to decide whether a given connected graph is Eulerian and to find an Euler circuit if the answer is affirmative.

B. Method

Here is a good characterization for Eulerian graphs:

> *A connected undirected graph is Eulerian if and only if it has no node of an odd degree, and a directed graph has an Euler circuit if and only if the number of incoming*

edges is equal to the number of outgoing edges at each node.

This characterization gives a straightforward procedure to decide whether a graph is Eulerian. Furthermore, an Euler circuit in an Eulerian graph G of m edges can be determined by the following method.

STEP 1. Choose any node u as the starting node, and traverse any edge (u, v) incident to node u, and then traverse any unused edge incident to node v. Repeat this process of traversing unused edges until the starting node u is reached. Let P be the resulting walk consisting of all used edges. If all edges of G are in P, then stop.

STEP 2. Choose any unused edge (x, y) in G such that x is in P and y is not in P. Use node x as the starting node and find another walk Q using all unused edges as in Step 1.

STEP 3. Walk P and walk Q share a common node x. They can be merged to form a walk R by starting at any node s of P and to traverse P until node x is reached; then, detour and traverse all edges of Q until node x is reached and continue to traverse the edges of P until the starting node s is reached. Set $P = R$.

STEP 4. Repeat Steps 2 and 3 until all edges are used.

The running time of the algorithm is $O(m)$.

C. Subroutine EULER parameters

Input:

N	Number of nodes.
M	Number of edges.
M2	Equal to M + M.
DIRECT	DIRECT = TRUE if the graph is directed; otherwise, DIRECT = FALSE.
INODE, JNODE	Each is an integer vector of length M; if DIRECT = FALSE then INODE(i) and

JNODE(i) are the two end nodes of the ith edge; if DIRECT = TRUE then the ith edge is directed from INODE(i) to JNODE(i); the input graph is assumed to be connected.

Output:

SUCESS	If SUCESS = TRUE then an Euler circuit is found; if SUCESS = FALSE then the input graph is not Eulerian.
TRAIL	Integer vector of length M; TRAIL(i) is the edge number of the ith edge in the Euler circuit, $i = 1, 2, \ldots, M$.

Working storages:

STACK	Integer vector of length M2; stack of edges.
ENDNOD	Integer vector of length M; storing the end nodes of the edges.
CANDID	Logical vector of length M; indicating whether the edges are next candidates.

D. Test example

Decide whether the following graph is Eulerian. If so, output an Euler circuit.

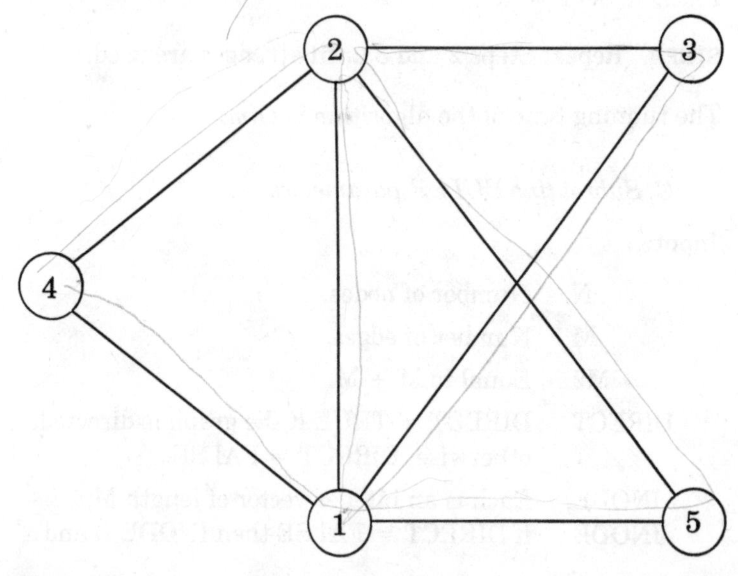

E. Program listing

MAIN PROGRAM

```
      INTEGER INODE(7),JNODE(7),TRAIL(7),
     +        STACK(14),ENDNOD(7)
      LOGICAL DIRECT,SUCESS,CANDID(7)
      DATA INODE /2,1,2,1,3,1,2/,
     +     JNODE /5,4,3,2,1,5,4/
C
      N = 5
      M = 7
      M2 = M + M
      DIRECT = .FALSE.
      CALL EULER (N,M,M2,DIRECT,INODE,
     +            JNODE,SUCESS,TRAIL,
     +            STACK,ENDNOD,CANDID)
      IF (SUCESS) THEN
        WRITE(*,10) (TRAIL(I),I=1,M)
 10     FORMAT(' THE EULER CIRCUIT IS:',
     +         //5X,7I4)
      ELSE
        WRITE(*,20)
 20     FORMAT(' THE INPUT GRAPH IS NOT',
     +         ' EULERIAN')
      ENDIF
      STOP
      END
```

OUTPUT RESULTS

```
THE EULER CIRCUIT IS:
1 6 5 3 7 2 4
```

```
      SUBROUTINE EULER(N,M,M2,DIRECT,
     +                 INODE,JNODE,
     +                 SUCESS,TRAIL,
     +                 STACK,ENDNOD,
```

```
     +                                    CANDID)
C
C          Decide whether a graph is Eulerian
C
           INTEGER INODE(M),JNODE(M),TRAIL(M),
     +              STACK(M2),ENDNOD(M)
           LOGICAL DIRECT,SUCESS,CANDID(M)
C
           DO 10 I = 1, N
              TRAIL(I) = 0
              ENDNOD(I) = 0
  10       CONTINUE
C
           IF (DIRECT) THEN
C
C             Check if the directed graph is Eulerian
C
              DO 20 I = 1, M
                 J = INODE(I)
                 TRAIL(J) = TRAIL(J) + 1
                 J = JNODE(I)
                 ENDNOD(J) = ENDNOD(J) + 1
  20          CONTINUE
              DO 30 I = 1, N
                 IF (TRAIL(I) .NE. ENDNOD(I)) THEN
                    SUCESS = .FALSE.
                    RETURN
                 ENDIF
  30          CONTINUE
           ELSE
C
C             Check if the undirected graph is Eulerian
C
              DO 40 I = 1, M
                 J = INODE(I)
                 ENDNOD(J) = ENDNOD(J) + 1
                 J = JNODE(I)
                 ENDNOD(J) = ENDNOD(J) + 1
  40          CONTINUE
```

```
            DO 50 I = 1, N
               IF (MOD(ENDNOD(I),2) .NE. 0) THEN
                  SUCESS = .FALSE.
                  RETURN
               ENDIF
   50       CONTINUE
         ENDIF
C
C        The input graph is Eulerian; find an Euler circuit
C
         SUCESS = .TRUE.
         LENSOL = 1
         LENSTK = 0
C
C        Find the next edge
C
   60    IF (LENSOL .EQ. 1) THEN
            ENDNOD(1) = JNODE(1)
            STACK(1) = 1
            STACK(2) = 1
            LENSTK = 2
         ELSE
            L = LENSOL - 1
            IF (LENSOL .NE. 2) ENDNOD (L)
     +         = INODE(TRAIL(L)) + JNODE(TRAIL(L))
     +         - ENDNOD (L - 1)
            K = ENDNOD(L)
            IF (DIRECT) THEN
               DO 70 I = 1, M
   70             CANDID(I) = K .EQ. INODE(I)
            ELSE
               DO 80 I = 1, M
   80             CANDID(I) = (K .EQ. INODE(I)) .OR.
     +               (K .EQ. JNODE(I))
            ENDIF
            DO 90 I = 1, L
   90          CANDID(TRAIL(I)) = .FALSE.
            LEN = LENSTK
            DO 100 I = 1, M
```

```
          IF (CANDID(I)) THEN
            LEN = LEN + 1
              STACK(LEN) = I
          ENDIF
 100    CONTINUE
        STACL(LEN + 1) = LEN - LENSTK
        LENSTK = LEN + 1
      ENDIF
C
C     Search further
C
 110  ISTAK = STACK(LENSTK)
      LENSTK = LENSTK - 1
      IF (ISTAK .EQ. 0) THEN
        LENSOL = LENSOL - 1
        IF (LENSOL .NE. 0) GOTO 110
        RETURN
      ELSE
        TRAIL(LENSOL) = STACK(LENSTK)
        STACK(LENSTK) = ISTAK - 1
        IF (LENSOL .EQ. M) RETURN
        LENSOL = LENSOL + 1
        GOTO 60
      ENDIF
C
      END
```

3-2 Hamiltonian Cycle

A. Problem description

A *Hamiltonian cycle* in a graph G is a cycle containing every node of G, and a graph is *Hamiltonian* if it has a Hamiltonian cycle. In the case of a directed graph, the cycle is directed. The problem is to decide whether a given graph is

Hamiltonian and to find a Hamiltonian cycle if the answer is affirmative.

B. Method

A Hamiltonian cycle in a given graph G of n nodes will be found by an exhaustive search. Start with a single node, say node 1, as the partially constructed cycle. The cycle is grown by a backtracking procedure until a Hamiltonian cycle is formed. More precisely, let

$$h(1), h(2), \ldots, h(k-1)$$

be the partially constructed cycle at the kth stage. Then, the set of candidates for the next element $h(k)$ is the set of all nodes u in G such that

if $k = 1$, then $u = 1$;
if $k > 1$, then $(h(k-1), u)$ is an edge in G, and u is distinct from $h(1), h(2), \ldots, h(k-1)$; furthermore,
if $k = n$, then $(u, h(1))$ is an edge in G and $u < h(2)$.

The search is complete when $k = n$. If the set of candidates is empty at any stage, then the graph is not Hamiltonian.

C. Subroutine HAMCYC parameters

Input:

N	Number of nodes.
M	Number of edges.
N1	Equal to N + 1.
DIRECT	DIRECT = TRUE if the graph is directed; otherwise; DIRECT = FALSE.
M2	If DIRECT = TRUE, then M2 = M; if DIRECT = FALSE, then M2 = M + M.
INODE,	Each is an integer vector of length M;

JNODE	if DIRECT = FALSE, then INODE(i) and JNODE(i) are the two end nodes of the ith edge; if DIRECT = TRUE, then the ith edge is directed from INODE(i) to JNODE(i).

Output:

SUCESS	If SUCESS = TRUE, then a Hamiltonian cycle is found; if SUCESS = FALSE, then the input graph is not Hamiltonian.
HCYCLE	Integer vector of length N; the nodes of the Hamiltonian cycle are stored in HCYCLE(i), $i = 1, 2, \ldots, N$.

Working storages:

FWDARC	Integer vector of length M2; FWDARC(i) is the ending node of the ith edge in the forward star representation of the graph.
ARCFIR	Integer vector of length N1; ARCFIR(i) is the number of the first edge starting at node i in the forward star representation of the graph.
STACK	Integer vector of length M2; stack of nodes.
CONECT	Logical vector of length N; indicating whether the node is connected to the edge being considered.

D. Test example

Decide whether the following graph is Hamiltonian. If so, output a Hamiltonian cycle.

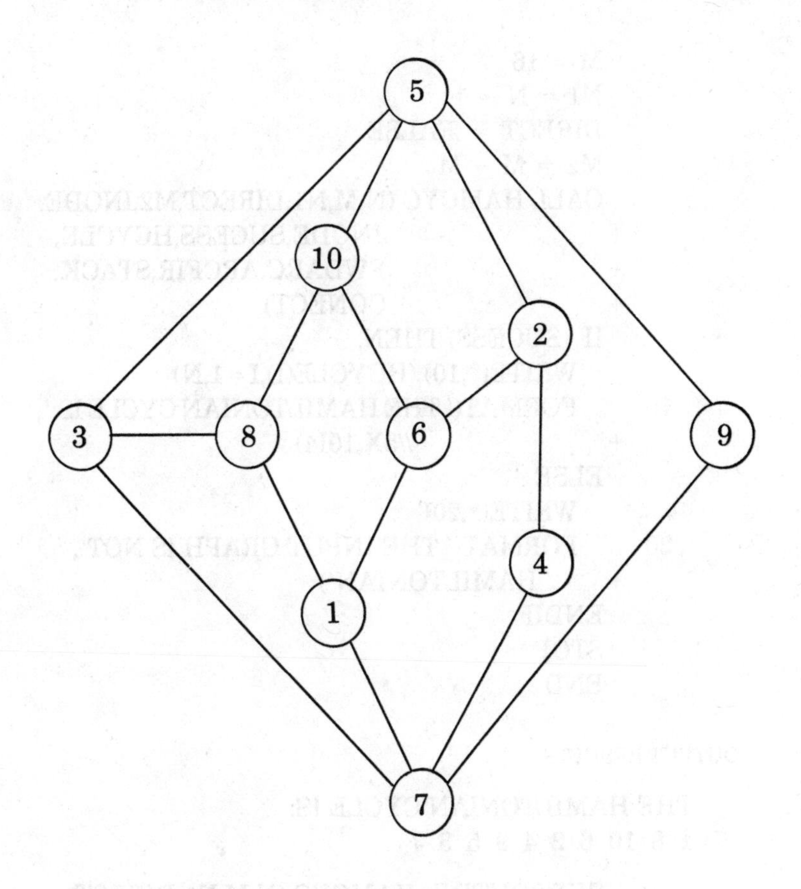

E. Program listing

MAIN PROGRAM

```
      INTEGER INODE(16),JNODE(16),
     +        HCYCLE(10),FWDARC(32),
     +        ARCFIR(11),
     +        STACK(32)
      LOGICAL DIRECT,SUCESS,CONECT(10)
      DATA INODE /2,3, 8,5,4,8,10,7,5,2,1,3,9,3, 6,6/,
     +     JNODE /6,5,10,2,9,1, 5,4,9,4,7,8,7,7,10,1/
C
      N = 10
```

```
      M = 16
      N1 = N + 1
      DIRECT = .FALSE.
      M2 = M + M
      CALL HAMCYC (N,M,N1,DIRECT,M2,INODE,
     +                     JNODE,SUCESS,HCYCLE,
     +                     FWDARC,ARCFIR,STACK,
     +                     CONECT)
      IF (SUCESS) THEN
        WRITE(*,10) (HCYCLE(I),I=1,N)
   10   FORMAT(' THE HAMILTONIAN CYCLE IS:',
     +                 //5X,10I4)
      ELSE
        WRITE(*,20)
   20   FORMAT(' THE INPUT GRAPH IS NOT',
     +     ' HAMILTONIAN')
      ENDIF
      STOP
      END
```

OUTPUT RESULTS

```
THE HAMILTONIAN CYCLE IS:
1 8 10 6 2 4 9 5 3 7
```

```
      SUBROUTINE HAMCYC (N,M,N1,DIRECT,
     +                          M2,INODE,JNODE,
     +                          SUCESS,HCYCLE,
     +                          FWDARC,ARCFIR,
     +                          STACK,CONECT)
C
C     Decide whether a graph is Hamiltonian
C
      INTEGER INODE(M),JNODE(M),
     +              HCYCLE(N),FWDARC(M2),
     +              ARCFIR(N1),
     +              STACK(M2)
      LOGICAL DIRECT,SUCESS,CONECT(N),JOIN
C
```

```
C       Set up the forward star representation of the
C          graph
C
        K = 0
        DO 20 I = 1, N
           ARCFIR(I) = K + 1
           DO 10 J = 1, M
              IF (INODE(J) .EQ. I) THEN
                 K = K + 1
                 FWDARC(K) = JNODE(J)
              ENDIF
              IF (.NOT. DIRECT) THEN
                 IF (JNODE(J) .EQ. I) THEN
                    K = K + 1
                    FWDARC(K) = INODE(J)
                 ENDIF
              ENDIF
 10        CONTINUE
 20     CONTINUE
        ARCFIR(N1) = K + 1
C
C       Initialize
C
        LENSOL = 1
        LENSTK = 0
C
C       Find the next node
C
 30     IF (LENSOL .EQ. 1) THEN
           STACK(1) = 1
           STACK(2) = 1
           LENSTK = 2
        ELSE
           LEN1 = LENSOL - 1
           LEN2 = HCYCLE(LEN1)
           DO 50 I = 1, N
              CONECT(I) = .FALSE.
              LOW = ARCFIR(LEN2)
```

```
            IUP = ARCFIR(LEN2 + 1)
            IF (IUP .GT. LOW) THEN
              IUP = IUP - 1
              DO 40 K = LOW, IUP
                IF (FWDARC(K) .EQ. I) THEN
                  CONECT(I) = .TRUE.
                  GOTO 50
                ENDIF
40            CONTINUE
            ENDIF
50        CONTINUE
          DO 60 I = 1, LEN1
            LEN = HCYCLE(I)
            CONECT(LEN) = .FALSE.
60        CONTINUE
          LEN = LENSTK
          IF (LENSOL .NE. N) THEN
            DO 70 I = 1, N
              IF (CONECT(I)) THEN
                LEN = LEN + 1
                STACK(LEN) = I
              ENDIF
70          CONTINUE
            STACK(LEN + 1) = LEN - LENSTK
            LENSTK = LEN + 1
          ELSE
            DO 100 I = 1, N
              IF (CONECT(I)) THEN
                IF (.NOT. DIRECT) THEN
                  IF (I .GT. HCYCLE(2)) THEN
                    STACK(LEN + 1) = LEN
     +                - LENSTK
                    LENSTK = LEN + 1
                    GOTO 110
                  ENDIF
                ENDIF
                JOIN = .FALSE.
                LOW = ARCFIR(I)
                IUP = ARCFIR(I + 1)
```

```
                  IF (IUP .GT. LOW) THEN
                    IUP = IUP - 1
                    DO 80 K = LOW, IUP
                      IF (FWDARC(K) .EQ. 1) THEN
                        JOIN = .TRUE.
                        GOTO 90
                      ENDIF
 80                 CONTINUE
                  ENDIF
 90               IF (JOIN) THEN
                    LENSTK = LENSTK + 2
                    STACK(LENSTK - 1) = I
                    STACK(LENSTK) = 1
                  ELSE
                    STACK(LEN + 1)
     +                = LEN - LENSTK
                    LENSTK = LEN + 1
                  ENDIF
                  GOTO 110
                ENDIF
100           CONTINUE
              STACK(LEN + 1) = LEN - LENSTK
              LENSTK = LEN + 1
            ENDIF
          ENDIF
C
C         Search further
C
110       ISTAK = STACK(LENSTK)
          LENSTK = LENSTK - 1
          IF (ISTAK .EQ. 0) THEN
            LENSOL = LENSOL - 1
            IF (LENSOL .EQ. 0) THEN
              SUCESS = .FALSE.
              RETURN
            ENDIF
            GOTO 110
          ELSE
            HCYCLE(LENSOL) = STACK(LENSTK)
```

```
        STACK(LENSTK) = ISTAK - 1
        IF (LENSOL .EQ. N) THEN
          SUCESS = .TRUE.
          RETURN
        ENDIF
        LENSOL = LENSOL + 1
        GOTO 30
      ENDIF
C
      END
```

4

NODE COLORING

4-1 Chromatic Number

A. Problem description

A *coloring* of an undirected graph G is an assignment of colors to nodes of G such that no adjacent nodes of G have the same color. A graph is *k-colorable* if there is a coloring of G using k colors. The *chromatic number of* G is the minimum k for which G is k-colorable. The problem is to find the chromatic number of a given graph.

B. Method

The coloring is done by a simple implicit enumeration tree search method. Initially, node 1 is assigned color 1, and the remaining nodes are colored sequentially so that node i is colored with the lowest-numbered color which has not been used so far to color any nodes adjacent to node i.

Let p be the number of colors required by this feasible coloring. Attempt to generate a feasible coloring using $q < p$ colors. To accomplish this, all nodes colored with p must be recolored. Thus, a backtrack step can be taken up to node u, where node $u + 1$ is the lowest index assigned color p. Attempt to color node u with its smallest feasible alternative color greater than its current color. If there is no such alternative color which is smaller than p, then backtrack to node $u - 1$. Otherwise, recolor node u and proceed forward, sequentially recoloring all nodes $u + 1, u + 2, \ldots,$ with the smallest feasible color until either node n is colored or some node v is reached which requires color p. In the former case, an improved coloring using q colors has been found; in this case, backtrack and attempt to find a better coloring using less than q colors. In the latter case, backtrack from node v and proceed forward as before.

The algorithm terminates when backtracking reaches node 1.

C. Subroutine VCOLOR parameters

Input:

N	Number of nodes.
M	Number of edges, M > 1.
N1	Equal to N + 1.
M2	Equal to M + M.
INODE, JNODE	Each is an integer vector of length M; INODE(i) and JNODE(i) are the end nodes of the ith edge in the graph; the input graph may have more than one component.

Output:

NCOLOR	The chromatic number of the graph.

COLOR	Integer vector of length N; COLOR(i) is the color assigned to node i.

Working storages:

FWDARC	Integer vector of length M2; FWDARC(i) is the ending node of the ith edge in the forward star representation of the graph.
ARCFIR	Integer vector of length N1; ARCFIR(i) is the number of the first edge starting at node i in the forward star representation of the graph.
DEGREE	Integer vector of length N; DEGREE(i) is the degree of node i.
CMAX1	Integer vector of length N; CMAX1(i) is the number of colors that can be used to color node i.
CMAX2	Integer vector of length N; CMAX2(i) is the largest color number used by nodes before node i.
COLOR1	Integer vector of length N; COLOR1(i) is the current color of node i.
COLOR2	Integer vector of length N; COLOR2(i) is the number of feasible colors available for node i.
AVAILC	Integer matrix of dimension N by N; Auxiliary storage of the available colors for each node.

D Test example

Find an optimal coloring for the following graph.

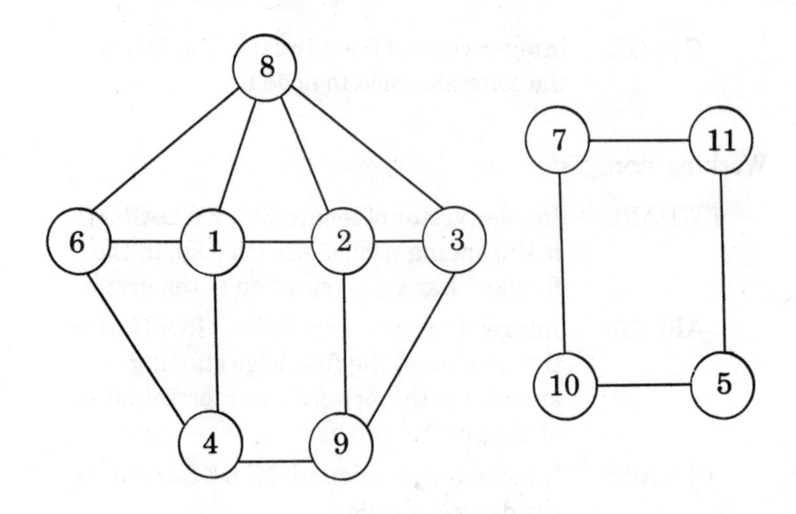

E. Program listing

MAIN PROGRAM

```
      INTEGER INODE(16),JNODE(16),COLOR(11),
     +        FWDARC(32),ARCFIR(12),
     +        DEGREE(11),CMAX1(11),
     +        CMAX2(11),COLOR1(11),
     +        COLOR2(11),
     +        AVAILC(11,11)
      DATA INODE /3,6, 7,1,4,10,2,6,8,
     +            5,8,10,1,3,2,9/,
     +     JNODE /9,8,11,2,1, 7,8,4,3,
     +            11,1, 5,6,2,9,4/
      N = 11
      M = 16
      N1 = N + 1
      M2 = M + M
      CALL VCOLOR (N,M,N1,M2,INODE,JNODE,
     +             NCOLOR,COLOR,
     +             FWDARC,ARCFIR,DEGREE,
     +             CMAX1,CMAX2,
     +             COLOR1,COLOR2,AVAILC)
      WRITE(*,10) NCOLOR
```

```
10    FORMAT(' THE NUMBER OF COLORS USED',
     +          ' = ',I3/)
      DO 20 I = 1, N
20      WRITE(*,30) I,COLOR(I)
30      FORMAT(' NODE',I3,' HAS COLOR',I3)
      STOP
      END
```

OUTPUT RESULTS

```
THE NUMBER OF COLORS USED = 4
      NODE  1 HAS COLOR 1
      NODE  2 HAS COLOR 2
      NODE  3 HAS COLOR 1
      NODE  4 HAS COLOR 2
      NODE  5 HAS COLOR 1
      NODE  6 HAS COLOR 3
      NODE  7 HAS COLOR 1
      NODE  8 HAS COLOR 4
      NODE  9 HAS COLOR 3
      NODE 10 HAS COLOR 2
      NODE 11 HAS COLOR 2
```

```
      SUBROUTINE VCOLOR (N,M,N1,M2,
     +                   INODE,JNODE,
     +                   NCOLOR,COLOR,
     +                   FWDARC,ARCFIR,
     +                   DEGREE,CMAX1,
     +                   CMAX2,COLOR1,
     +                   COLOR2,AVAILC)
C
C     Node coloring
C
      INTEGER INODE (M),JNODE(M),COLOR(N),
     +        FWDARC(M2),ARCFIR(N1),
     +        DEGREE(N),CMAX1(N),
     +        CMAX2(N),COLOR1(N),
```

```
     +              COLOR2(N),AVAILC(N,N)
      LOGICAL MORE
C
C     Set up the forward star representation of the
C       graph
C
      DO 10 I = 1, N
 10     DEGREE(I) = 0
      K = 0
      DO 30 I = 1, N
        ARCFIR(I) = K + 1
        DO 20 J = 1, M
          IF (INODE(J) .EQ. I) THEN
            K = K + 1
            FWDARC(K) = JNODE(J)
            DEGREE(I) = DEGREE(I) + 1
          ELSE
            IF (JNODE(J) .EQ. I) THEN
              K = K + 1
              FWDARC(K) = INODE(J)
              DEGREE(I) = DEGREE(I) + 1
            ENDIF
          ENDIF
 20     CONTINUE
 30   CONTINUE
      ARCFIR(N1) = K + 1
C
      DO 60 I = 1, N
        COLOR2(I) = DEGREE(I) + 1
        IF (COLOR2(I) .GT. I) COLOR2(I) = I
        CMAX1(I) = COLOR2(I)
        LOOP = COLOR2(I)
        DO 40 J = 1, LOOP
 40       AVAILC(I,J) = N
        K = COLOR2(I) + 1
        IF (K .LE. N) THEN
        DO 50 J = K, N
 50       AVAILC(I,J) = 0
```

```
        ENDIF
 60   CONTINUE
      NOD = 1
C
C     Color node NOD
C
      NEWC = 1
      NCOLOR = N
      IPAINT = 0
      MORE = .TRUE.
C
 70   IF (MORE) THEN
        INDEX = CMAX1(NOD)
        IF (INDEX .GT. IPAINT + 1) INDEX
     +     = IPAINT + 1
 80     IF ((AVAILC(NOD,NEWC) .LT. NOD) .AND.
     +      (NEWC .LE. INDEX)) THEN
          NEWC = NEWC + 1
          GOTO 80
        ENDIF
C
C       Node NOD has the color NEWC
C
        IF (NEWC .EQ. INDEX + 1) THEN
          MORE = .FALSE.
        ELSE
          IF (NOD .EQ. N) THEN
C
C           A new coloring is found
C
            COLOR1(NOD) = NEWC
            DO 90 I = 1, N
 90           COLOR(I) = COLOR1(I)
            IF (NEWC .GT. IPAINT) IPAINT
     +         = IPAINT + 1
            NCOLOR = IPAINT
            IF (NCOLOR .GT. 2) THEN
C
```

```
C           Backtrack to the first node of color
C             NCOLOR
C
            INDEX = 1
 100        IF (COLOR(INDEX) .NE. NCOLOR)
     +        THEN
              INDEX = INDEX + 1
              GOTO 100
            ENDIF
            J = N
 110        IF (J .GE. INDEX) THEN
              NOD = NOD - 1
              NEWC = COLOR1(NOD)
              IPAINT = CMAX2(NOD)
              LOW = ARCFIR(NOD)
              IUP = ARCFIR(NOD + 1)
              IF (IUP .GT. LOW) THEN
                IUP = IUP - 1
                DO 120 K = LOW, IUP
                  NODEK = FWDARC(K)
                  IF (NODEK .GT. NOD) THEN
                    IF (AVAILC(NODEK,NEWC)
     +                .EQ. NOD) THEN
                      AVAILC(NODEK,NEWC)
     +                  = N
                      COLOR2(NODEK)
     +                  = COLOR2(NODEK) + 1
                    ENDIF
                  ENDIF
 120            CONTINUE
              ENDIF
              NEWC = NEWC + 1
              MORE = .FALSE.
              J = J - 1
              GOTO 110
            ENDIF
            IPAINT = NCOLOR - 1
            DO 140 I = 1, N
              LOOP = CMAX1(I)
```

```
                IF (LOOP .GT. IPAINT) THEN
                  K = IPAINT + 1
                  DO 130 J = K, LOOP
                    IF (AVAILC(I,J) .EQ. N)
     +                COLOR2(I) = COLOR2(I) - 1
130               CONTINUE
                  CMAX1(I) = IPAINT
                ENDIF
140           CONTINUE
            ENDIF
          ELSE
C
C           NOD is less than N
C
            LOW = ARCFIR(NOD)
            IUP = ARCFIR(NOD + 1)
            IF (IUP .GT. LOW) THEN
              IUP = IUP - 1
              K = LOW
150           IF ((K .LE. IUP) .AND. MORE) THEN
                NODEK = FWDARC (K)
                IF (NODEK .GT. NOD)
     +            MORE = .NOT.
     +              ((COLOR2(NODEK) .EQ. 1)
     +              .AND.
     +              (AVAILC(NODEK,NEWC)
     +                .GE. NOD))
                K = K + 1
                GOTO 150
              ENDIF
            ENDIF
C
            IF (MORE) THEN
              COLOR1(NOD) = NEWC
              CMAX2(NOD) = IPAINT
              IF (NEWC .GT. IPAINT)
     +          IPAINT = IPAINT + 1
              LOW = ARCFIR(NOD)
              IUP = ARCFIR(NOD + 1)
```

```
            IF (IUP .GT. LOW) THEN
              IUP = IUP - 1
              DO 160 K = LOW,IUP
                NODEK = FWDARC(K)
                IF (NODEK .GT. NOD) THEN
                  IF (AVAILC(NODEK,NEWC)
     +                .GE. NOD) THEN
                    AVAILC(NODEK,NEWC)
     +                  = NOD
                    COLOR2(NODEK)
     +                  = COLOR2(NODEK) - 1
                  ENDIF
                ENDIF
160           CONTINUE
            ENDIF
            NOD = NOD + 1
            NEWC = 1
          ELSE
            NEWC = NEWC + 1
          ENDIF
        ENDIF
      ENDIF
    ELSE
      MORE = .TRUE.
      IF ((NEWC .GT. CMAX1(NOD)) .OR.
     +    (NEWC .GT. IPAINT + 1)) THEN
        NOD = NOD - 1
        NEWC = COLOR1(NOD)
        IPAINT = CMAX2(NOD)
        LOW = ARCFIR(NOD)
        IUP = ARCFIR(NOD + 1)
        IF (IUP .GT. LOW) THEN
          IUP = IUP - 1
          DO 170 K = LOW, IUP
            NODEK = FWDARC(K)
            IF (NODEK .GT. NOD) THEN
              IF (AVAILC(NODEK,NEWC) .EQ.
     +          NOD) THEN
                AVAILC(NODEK,NEWC) = N
```

```
                   COLOR2(NODEK)
     +                  = COLOR2(NODEK) + 1
                 ENDIF
               ENDIF
 170         CONTINUE
           ENDIF
           NEWC = NEWC + 1
           MORE = .FALSE.
         ENDIF
       ENDIF
       IF ((NOD .NE. 1) .AND. (NCOLOR .NE. 2))
     +   GOTO 70
C
       RETURN
       END
```

4-2 Chromatic Polynomial

A. Problem description

Let $f(p;G)$ be the number of colorings of a connected graph G in p or fewer colors. For any graph G with n nodes, $f(p;G)$ is a polynomial in p. The *chromatic polynomial* $f(p;G)$ can be expressed in three common forms:

1)

$$f(p;G) = a_n p^n - a_{n-1}p^{n-1} + a_{n-2}p^{n-2} - \ldots + (-1)^{n-1}a_1 p$$

in which the coefficients $a_i \geq 0$, $i = 1, 2, \ldots, n-1$ alternate in sign, and $a_n = 1$.

2)

$$f(p;G) = \sum_{i=1}^{n} (-1)^{n-i} b_i p(p-1)^{i-1}$$

where $b_i \geq 0$, $i = 1, 2, \ldots, n$.

3)

$$f(p;G) = \sum_{i=1}^{n} c_i p_{(i)}$$

where $p_{(i)} = p(p-1)(p-2)\ldots(p-i+1)$.

The problem is to find the chromatic polynomial of a given connected graph.

B. Method

Consider two adjacent nodes u, v in a graph G. Let H be the graph obtained from G by deleting the edge (u, v) from G, and let F be the graph obtained from G by merging the two nodes u, v in G. Thus,

$$f(p;G) = f(p;H) - f(p;F).$$

The chromatic polynomial $f(p;G)$ can be obtained by repeated application of this reduction formula, thereby expressing $f(p;G)$ as a linear combination of chromatic polynomials of null graphs. The complete derivation can be represented by a binary tree in which left and right branches correspond to deletion and merging steps, respectively.

The running time of the algorithm is $O(m \cdot 2^m)$, where m is the number of edges in the input graph.

C. Subroutine CHPOLY parameters

Input:

N	Number of nodes.
M	Number of edges.
NM	Equal to (N * (M + M − N + 1))/2
INODE, JNODE	Each is an integer vector of length M; INODE(i) and JNODE(i) are the end nodes of the ith edge in the graph, and the input graph is assumed to be connected.

Output:

CPOLY1	Integer vector of length N; coefficients in equation (1).
CPOLY2	Integer vector of length N; coefficients in equation (2).

CPOLY3 Integer vector of length N; coefficients in equation (3).

Working storages:

ISTACK, JSTACK Each is an integer vector of length NM; these are stacks that control the traversal of the binary tree.

D. Test example

Find the chromatic polynomial of the following graph.

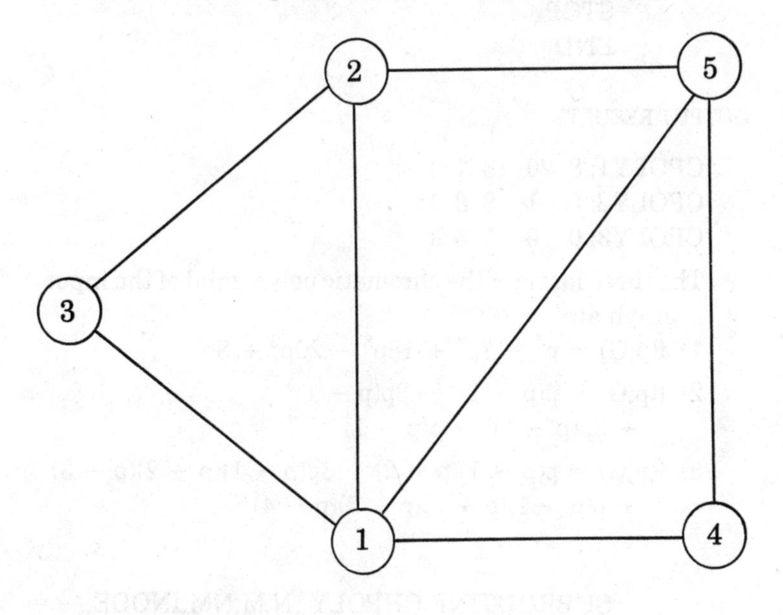

E. Program listing

MAIN PROGRAM

```
      INTEGER INODE(7),JNODE(7),CPOLY1(5),
     +        CPOLY2(5),CPOLY3(5),
     +        ISTACK(25),JSTACK(25)
      DATA INODE /1, 1, 5, 1, 3, 1, 5/,
     +     JNODE /2, 3, 2, 4, 2, 5, 4/
```

```
      N = 5
      M = 7
      NM = (N * (M + M - N + 1)) / 2
      CALL CHPOLY (N,M,NM,INODE,JNODE,
     +             CPOLY1,CPOLY2,CPOLY3,
     +             ISTACK,JSTACK)
      WRITE(*,10) (CPOLY1(I),I=1,N),
     +            (CPOLY2(I),I=1,N),
     +            (CPOLY3(I),I=1,N)
   10 FORMAT(' CPOLY1: ',5I5/' CPOLY2: ',5I5/
     +       'CPOLY3: ',5I5)
      STOP
      END
```

OUTPUT RESULTS

```
CPOLY1: 8 20 18 7 1
CPOLY2: 0  1  3 3 1
CPOLY3: 0  0  1 3 1
```

The three forms of the chromatic polynomial of the input graph are:

1) $f(p;G) = p^5 - 7p^4 + 18p^3 - 20p^2 + 8p$
2) $f(p;G) = p(p-1)^4 - 3p(p-1)^3 + 3p(p-1)^2 - p(p-1)$
3) $f(p;G) = p(p-1)(p-2) + 3p(p-1)(p-2)(p-3) + p(p-1)(p-2)(p-3)(p-4)$.

```
      SUBROUTINE CHPOLY (N,M,NM,INODE,
     +                   JNODE,CPOLY1,
     +                   CPOLY2,CPOLY3,
     +                   ISTACK,JSTACK)
C
C     Chromatic polynomial
C
      INTEGER INODE(M),JNODE(M),CPOLY1(N)
     +        CPOLY2(N),CPOLY3(N),
     +        ISTACK(NM),JSTACK(NM)
```

```
      LOGICAL VISIT,NONPOS
C
      ITOP = 0
      DO 10 I = 1, N
 10     CPOLY2(I) = 0
      MM = M
      NN = N
C
C     Find a spanning tree
C
 20   MAXMN = MM + 1
      IF (NN .GT. MM) MAXMN = NN + 1
      DO 30 I = 1, NN
 30     CPOLY3(I) = -I
      DO 40 I = 1, MM
        J = INODE(I)
        INODE(I) = CPOLY3(J)
        CPOLY3(J) = - MAXMN - I
        J = JNODE(I)
        JNODE(I) = CPOLY3(J)
        CPOLY3(J) = - MAXMN - MAXMN - I
 40   CONTINUE
C
      NCOMP = 0
      INDEX = 0
      NODEW = 0
 50   NODEW = NODEW + 1
      IF (NODEW .LE. NN) THEN
        NODEV = CPOLY3(NODEW)
        IF (NODEV .GT. 0) GOTO 50
        NCOMP = NCOMP + 1
        CPOLY3(NODEW) = NCOMP
        IF (NODEV .GE. -NODEW) GOTO 50
        NODEU = NODEW
        NODEX = -NODEV
        VISIT = .TRUE.
        ISUB2 = -NODEV / MAXMN
        JSUB2 = -NODEV - ISUB2 * MAXMN
```

```
60      IF (ISUB2 .EQ. 1) THEN
          NODEV = INODE(JSUB2)
        ELSE
          NODEV = JNODE(JSUB2)
        ENDIF
        IF (NODEV .GT. 0) THEN
          IF (NODEV .LE. MAXMN) THEN
            IF (CPOLY3(NODEV) .GE. 0) NODEY
   +          = JSUB2
          ENDIF
        ENDIF
        IF (NODEV .GE. 0) THEN
          IF (IX .EQ. 1) THEN
            NODEX = IABS(INODE(IY))
            INODE(IY) = NODEU
          ELSE
            NODEX = IABS(JNODE(IY))
            JNODE(IY) = NODEU
          ENDIF
          IX = NODEX / MAXMN
          IY = NODEX - IX * MAXMN
          IF (IX .EQ. 0) THEN
70            IF (NODEY .NE. 0) THEN
                INDEX = INDEX + 1
                CPOLY3(NODEU) = INODE(NODEY)
   +              + JNODE(NODEY) - NODEU
                INODE(NODEY) = -INDEX
                JNODE(NODEY) = NODEU
              ENDIF
            NODEU = IY
            IF (IY .LE. 0) GOTO 50
            NODEX = -CPOLY3(NODEU)
            IX = NODEX / MAXMN
            IY = NODEX - IX * MAXMN
            IF (IX .EQ. 0) GOTO 70
          ENDIF
          ISUB2 = 3 - IX
```

```
        JSUB2 = IY
        GOTO 60
      ENDIF
C
      IF (NODEV .LT. -MAXMN) THEN
        IF (ISUB2 .EQ. 1) THEN
          INODE(JSUB2) = -NODEV
        ELSE
          JNODE(JSUB2) = -NODEV
        ENDIF
        ISUB2 = -NODEV / MAXMN
        JSUB2 = -NODEV - ISUB2 * MAXMN
      ELSE
        NODEV = - NODEV
        IF (ISUB2 .EQ. 1) THEN
          INODE(JSUB2) = 0
        ELSE
          JNODE(JSUB2) = 0
        ENDIF
        IF (VISIT) THEN
          ISUB1 = ISUB2
          JSUB1 = JSUB2
          NODEY = 0
          VISIT = .FALSE.
        ELSE
          IF (ISUB1 .EQ. 1) THEN
            INODE(JSUB1) = NODEV
          ELSE
            JNODE(JSUB1) = NODEV
          ENDIF
          ISUB1 = ISUB2
          JSUB1 = JSUB2
          IF (IX .EQ. 1) THEN
            NODEX = IABS(INODE(IY))
            INODE(IY) = NODEU
          ELSE
            NODEX = IABS(JNODE(IY))
```

```
               JNODE(IY) = NODEU
             ENDIF
           ENDIF
           IX = NODEX / MAXMN
           IY = NODEX - IX * MAXMN
           IF (IX .EQ. 0) THEN
   80        IF (NODEY .NE. 0) THEN
               INDEX = IDEX + 1
               CPOLY3(NODEU) = INODE(NODEY)
     +           + JNODE(NODEY) - NODEU
               INODE(NODEY) = -INDEX
               JNODE(NODEY) = NODEU
             ENDIF
             NODEU = IY
             IF (IY .LE. 0) GOTO 50
             NODEX = -CPOLY3(NODEU)
             IX = NODEX / MAXMN
             IY = NODEX - IX * MAXMN
             IF (IX .EQ. 0) GOTO 80
           ENDIF
           ISUB2 = 3 - IX
           JSUB2 = IY
         ENDIF
         GOTO 60
       ENDIF
C
       DO 100 I = 1, MM
   90    NODEY = -INODE(I)
         IF (NODEY .GE. 0) THEN
           NODEX = JNODE(I)
           JNODE(I) = JNODE(NODEY)
           JNODE(NODEY) = NODEX
           INODE(I) = INODE(NODEY)
           INODE(NODEY)
     +       = CPOLY3(JNODE(NODEY))
           GOTO 90
         ENDIF
  100  CONTINUE
C
```

```
          DO 110 I = 1, INDEX
110         CPOLY3(JNODE(I)) = CPOLY3(INODE(I))
C
C         If NCOMP is not equal to 1, then the graph is
C           not connected
C
          IF (NCOMP .NE. 1) THEN
120         DO 130 I = 1, N
              CPOLY1(I) = CPOLY2(I)
              CPOLY3(I) = CPOLY2(I) *
     +          (1 - 2 * MOD(N-I,2))
130         CONTINUE
C
            DO 150 I = 1, N
              JVERTX = 0
              DO 140 J = I, N
                JVERTX = CPOLY1(N + I - J)
     +            + JVERTX
                CPOLY1(N + I - J) = JVERTX
140           CONTINUE
150         CONTINUE
            INCR = 0
            DO 170 I = 1, N
              JVERTX = 0
              DO 160 J = I, N
                JVERTX = CPOLY3(N + I - J)
     +            + INCR * JVERTX
                CPOLY3(N + I - J) = JVERTX
160           CONTINUE
              INCR = INCR + 1
170         CONTINUE
            RETURN
          ENDIF
          IF (MM .LT. NN) THEN
            CPOLY2(NN) = CPOLY2(NN) + 1
            IF (ITOP .EQ. 0) GOTO 120
            NN = ISTACK(ITOP)
            MM = JSTACK(ITOP)
            ITOP = ITOP - MM - 1
```

```
          DO 180 I = 1, MM
            INODE(I) = ISTACK(ITOP + I)
            JNODE(I) = JSTACK(ITOP + I)
180       CONTINUE
          IF (MM .EQ. NN) THEN
            CPOLY2(NN) = CPOLY2(NN) + 1
          ELSE
            ITOP = ITOP + MM
            ISTACK(ITOP) = NN
            JSTACK(ITOP) = MM - 1
          ENDIF
        ELSE
          IF (MM .EQ. NN) THEN
            CPOLY2(NN) = CPOLY2(NN) + 1
          ELSE
            DO 190 I = 1, MM
              ITOP = ITOP + 1
              ISTACK(ITOP) = INODE(I)
              JSTACK(ITOP) = JNODE(I)
190         CONTINUE
            ISTACK(ITOP) = NN
            JSTACK(ITOP) = MM - 1
          ENDIF
        ENDIF
        DO 200 I = 1, N
200       CPOLY1(I) = 0
        IF (INODE(MM) .LT. JNODE(MM)) THEN
          IVERTX = INODE(MM)
        ELSE
          IVERTX = JNODE(MM)
        ENDIF
        JVERTX = INODE(MM) + JNODE(MM)
     +    - IVERTX
        LOOP = MM - 1
        MM = 0
        DO 210 I = 1, LOOP
          ILAST = INODE(I)
          IF (ILAST .EQ. JVERTX) ILAST = IVERTX
          IF (ILAST .EQ. NN) ILAST = JVERTX
```

```
         JLAST = JNODE(I)
         IF (JLAST .EQ. JVERTX) JLAST = IVERTX
         IF (JLAST .EQ. NN) JLAST = JVERTX
         IF (ILAST .EQ. IVERTX) THEN
           IF (CPOLY1(JLAST) .NE. 0) GOTO 210
           CPOLY1(JLAST) = 1
         ENDIF
         IF (JLAST .EQ. IVERTX) THEN
           IF (CPOLY1(ILAST) .NE. 0) GOTO 210
           CPOLY1(ILAST) = 1
         ENDIF
         MM = MM + 1
         INODE(MM) = ILAST
         JNODE(MM) = JLAST
210    CONTINUE
       NN = NN - 1
       GOTO 20
C
       END
```

5

MINIMUM SPANNING TREE

A. Problem description

Consider an undirected graph G with given edge lengths. The *minimum spanning tree problem* is to find a spanning tree in G such that the sum of the edge lengths in the tree is minimum.

B. Method 1

Choose any node j and let this single node j be the partially constructed tree T. Join to T an edge whose length is minimal among all edges with one end in T and the other end not in T. Repeat this process until T becomes a spanning tree of G.

The running time of the algorithm is $O(n^2)$, where n is the number of nodes in the graph. The subroutine MINTR1 below implements this procedure.

Method 2

Initially, the set T is empty. Edges are considered for inclusion in T in the nondecreasing order of their lengths. An edge is included in T if it does not form a cycle with the edges already in T. A minimum spanning tree is formed when $n - 1$ edges are included in T.

In the implementation, the edges are partially sorted with the smallest edge at the root of a heap structure (a binary tree in which the weight of every node is not greater than the weights of its sons).

The running time of the algorithm is $O(m \log m)$, where m is the number of edges in the graph. The subroutine MINTR2 implements this procedure.

C. Subroutine MINTR1 parameters

Input:

N	Number of nodes.
DIST	Real symmetric matrix of dimension N by N; DIST(i, j) is the length of the edge between node i and node j, DIST(i, i) = 0 for all i, and DIST(i, j) = BIG if edge (i, j) is not in the graph.
NDIM	Row dimension of matrix DIST exactly as specified in the dimension statement of the calling program.
BIG	A sufficiently large real number greater than the sum of all edge lengths in the graph.

Output:

MTREE	Integer vector of length N; the ith edge of the minimum spanning tree is (i, MTREE(i)), i = 1, 2, . . . , N − 1.

Subroutine MINTR2 parameters

Input:

N	Number of nodes.
M	Number of edges.
INODE, JNODE	Each is an integer vector of length M; INODE(i) and JNODE(i) are the two end nodes of the ith edge of the input graph.
ARCLEN	Integer vector of length M; ARCLEN(i) is the length of the ith edge.

Output:

ITARC	If the input graph is connected, then ITARC will have the value N − 1; if the input graph is not connected, then ITARC will be the number of edges in the minimum spanning forest.
ITREE1, ITREE2	Each is an integer vector of length N; the edges in the minimum spanning tree or the minimum spanning forest are stored in: (ITREE1(i), ITREE2(i)), $i = 1, 2, \ldots,$ ITARC.

Working storages:

PRED	Integer vector of length N; predecessor array.

D. Test example

Find a minimum spanning tree of the following graph.

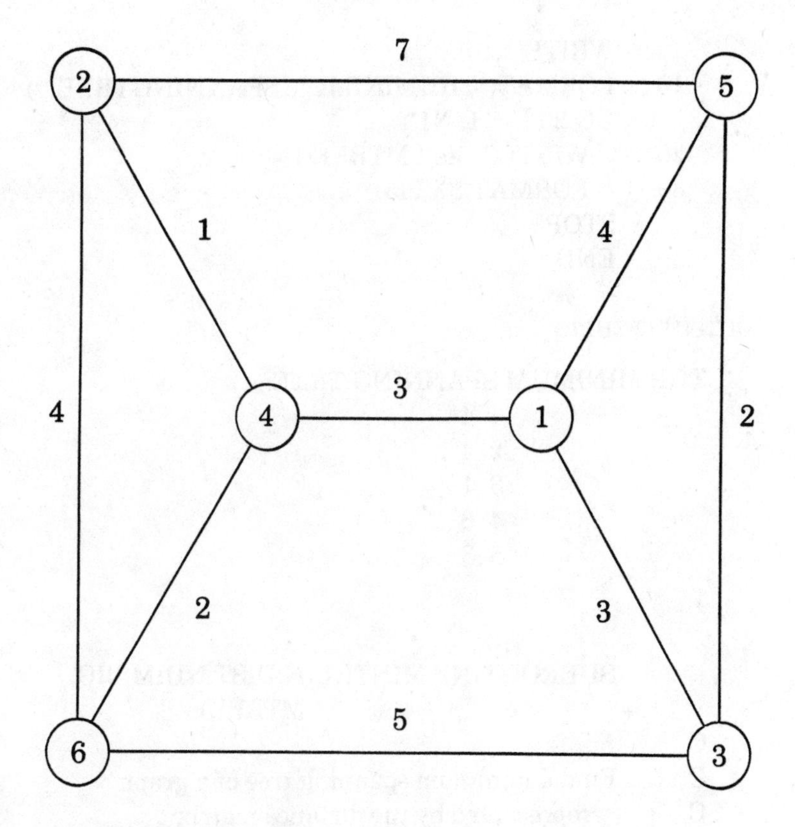

E. Program listing

MAIN PROGRAM (METHOD 1)

```
      REAL DIST(6,6)
      INTEGER MTREE(6)
      DATA DIST/ 0.,99.,3.,3.,4.,99., 99.,0.,99.,1.,7.,4.,
     +           3.,99.,0.,99.,2.,5., 3.,1.,99.,0.,99.,2.,
     +           4.,7.,2.,99.,0.,99., 99.,4.,5.,2.,99.,0./
C
      N = 6
      NDIM = 6
      BIG = 999.
      CALL MINTR1 (N,DIST,NDIM,BIG,MTREE)
      N1 = N - 1
```

```
      WRITE(*,10)
  10  FORMAT('THE MINIMUM SPANNING TREE:'/)
      DO 20 I = 1, N1
  20    WRITE(*,30) I,MTREE(I)
  30    FORMAT(8X,2I5)
      STOP
      END
```

OUTPUT RESULTS

```
  THE MINIMUM SPANNING TREE:
                 1 4
                 2 4
                 3 1
                 4 6
                 5 3
```

```
      SUBROUTINE MINTR1 (N,DIST,NDIM,BIG,
     +                       MTREE)
C
C     Find a minimum spanning tree of a graph
C       represented by the distance matrix
C
      REAL DIST(NDIM,N)
      INTEGER MTREE(N)
C
      N1 = N - 1
      DO 10 I = 1, N1
  10    MTREE(I) = -N
C
      MTREE(N) = 0
      DO 40 I = 1, N1
        DISTMN = BIG
        DO 20 J = 1, N1
          INODE = MTREE(J)
          IF (INODE .LE. 0) THEN
            D = DIST(-INODE,J)
            IF (D .LT. DISTMN) THEN
```

```
                DISTMN = D
                IMIN = J
              ENDIF
            ENDIF
 20       CONTINUE
          MTREE(IMIN) = -MTREE(IMIN)
          DO 30 J = 1, N1
            INODE = MTREE(J)
            IF (INODE .LE. 0) THEN
              IF (DIST(J,IMIN) .LT. DIST(J,-INODE))
     +          MTREE(J) = -IMIN
            ENDIF
 30       CONTINUE
 40     CONTINUE
C
        RETURN
        END
```

MAIN PROGRAM (METHOD 2)

```
        INTEGER INODE(9),JNODE(9),ARCLEN(9),
     +            ITREE1(6),ITREE2(6),
     +            PRED(6)
        DATA INODE /2,4,6,2,3,4,1,2,5/,
     +       JNODE /5,1,3,4,5,6,3,6,1/,
     +       ARCLEN /7,3,5,1,2,2,3,4,4/
C
        N = 6
        M = 9
        CALL MINTR2(N,M,INODE,JNODE,ARCLEN,
     +              ITARC,ITREE1,ITREE2,PRED)
        WRITE(*,10)
 10     FORMAT(' THE MINIMUM SPANNING',
     +         ' TREE:'/)
        DO 20 I = 1, ITARC
 20       WRITGE(*,30) ITREE1(I),ITREE2(I)
 30       FORMAT(8X,2I5)
        STOP
        END
```

OUTPUT RESULTS

```
THE MINIMUM SPANNING TREE:
            2 4
            3 5
            4 6
            4 1
            1 3
```

```
      SUBROUTINE MINTR2(N,M,INODE,JNODE,
     +                  ARCLEN,ITARC,
     +                  ITREE1,ITREE2,
     +                  PRED)
C
C     Find a minimum spanning tree of a graph
C       represented by a list of edges
C
      INTEGER INODE(M),JNODE(M),ARCLEN(M),
     +        ITREE1(N),ITREE2(N),
     +        PRED(N)
C
      DO 10 I = 1, N
 10     PRED(I) = -1
C
C     Initialize the heap structure
C
      I = M / 2
 20   IF (I .GT. 0) THEN
        INDEX1 = I
        MD2 = M / 2
 30     IF (INDEX1 .LE. MD2) THEN
          INDEX = INDEX1 + INDEX1
          IF ((INDEX .LT. M) .AND.
     +       (ARCLEN(INDEX + 1) .LT.
     +         ARCLEN(INDEX))) THEN
            INDEX2 = INDEX + 1
          ELSE
            INDEX2 = INDEX
          ENDIF
          IF (ARCLEN(INDEX2) .LT.
```

```
     +        ARCLEN(INDEX1)) THEN
              NODEU = INODE(INDEX1)
              NODEV = JNODE(INDEX1)
              LEN = ARCLEN(INDEX1)
              INODE(INDEX1) = INODE(INDEX2)
              JNODE(INDEX1) = JNODE(INDEX2)
              ARCLEN(INDEX1)
     +          = ARCLEN(INDEX2)
              INODE(INDEX2) = NODEU
              JNODE(INDEX2) = NODEV
              ARCLEN(INDEX2) = LEN
              INDEX1 = INDEX2
            ELSE
              INDEX1 = M
            ENDIF
            GOTO 30
          ENDIF
          I = I - 1
          GOTO 20
        ENDIF
        NEDGE = M
        ITARC = 0
        NUMARC = 0
        NM1 = N - 1
C
 40     IF ((ITARC .LT. NM1) .AND.
     +    (NUMARC .LT. M)) THEN
C
C         Examine the next edge
C
          NUMARC = NUMARC + 1
          NODEU = INODE(1)
          NODEV = JNODE(1)
          IPT = NODEU
C
C         Check if NODEU and NODEV are in the
C           same component
C
 50       IF (PRED(IPT) .GT. 0) THEN
```

```
            IPT = PRED(IPT)
            GOTO 50
          ENDIF
          INDEX1 = IPT
          IPT = NODEV
   60     IF (PRED(IPT) .GT. 0) THEN
            IPT = PRED(IPT)
            GOTO 60
          ENDIF
          INDEX2 = IPT
          IF (INDEX1 .NE. INDEX2) THEN
C
C           Include NODEU and NODEV in the
C             minimum spanning tree
C
            IPRED = PRED(INDEX1)
     +        + PRED(INDEX2)
            IF (PRED(INDEX1) .GT. PRED(INDEX2))
     +        THEN
              PRED(INDEX1) = INDEX2
              PRED(INDEX2) = IPRED
            ELSE
              PRED(INDEX2) = INDEX1
              PRED(INDEX1) = IPRED
            ENDIF
            ITARC = ITARC + 1
            ITREE1(ITARC) = NODEU
            ITREE2(ITARC) = NODEV
          ENDIF
C
C         Restore the heap structure
C
          INODE(1) = INODE(NEDGE)
          JNODE(1) = JNODE(NEDGE)
          ARCLEN(1) = ARCLEN(NEDGE)
          NEDGE = NEDGE - 1
          INDEX1 = 1
          NEDGE2 = NEDGE / 2
   70     IF (INDEX1 .LE. NEDGE2) THEN
```

```
          INDEX = INDEX1 + INDEX1
          IF ((INDEX .LT. NEDGE) .AND.
     +       (ARCLEN(INDEX + 1) .LT.
     +         ARCLEN(INDEX))) THEN
            INDEX2 = INDEX + 1
          ELSE
            INDEX2 = INDEX
          ENDIF
          IF (ARCLEN(INDEX2) .LT.
     +      ARCLEN(INDEX1)) THEN
            NODEU = INODE(INDEX1)
            NODEV = JNODE(INDEX1)
            LEN = ARCLEN(INDEX1)
            INODE(INDEX1) = INODE(INDEX2)
            JNODE(INDEX1) = JNODE(INDEX2)
            ARCLEN(INDEX1)
     +        = ARCLEN(INDEX2)
            INODE(INDEX2) = NODEU
            JNODE(INDEX2) = NODEV
            ARCLEN(INDEX2) = LEN
            INDEX1 = INDEX2
          ELSE
            INDEX1 = NEDGE
          ENDIF
          GOTO 70
        ENDIF
        GOTO 40
      ENDIF
C
      RETURN
      END
```

6

MAXIMUM CARDINALITY-MATCHING

A. Problem description

A *matching* in an undirected graph G is a subset S of edges of G in which no two edges in S are adjacent in G. The problem is to find a matching of maximum cardinality.

B. Method

Let S be a matching in an undirected graph G. A node that is not matched is called an *exposed* node. An *alternating path* with respect to a given matching S is a path in which the edges are alternately in S and not in S. An *augmenting path* is an alternating path which begins with an exposed node and ends with another exposed node. A fundamental theorem states that a matching S in G is maximum if and only if G has no augmenting path with respect to S.

The basic method for the maximum matching problem starts with an arbitrary matching Q. An augmeting path P

with respect to Q is found. Then, a new matching is constructed by taking those edges of Q or P that are not in both Q and P. The process is repeated and the matching is maximum when no augmenting path is found.

The whole algorithm runs $O(n^3)$ in time, where n is the number of nodes in the graph.

C. Subroutine MATCH parameters

Input:

N	Number of nodes.
M	Number of edges.
N1	Equal to N + 1.
M2	Equal to M + M.
INODE, JNODE	Each is an integer vector of length M; INODE(i) and JNODE(i) are the end nodes of the ith edge of the input graph which is not necessarily connected.

Output:

PAIR	Integer vector of length N; node i is matched with PAIR(i), $i = 1, 2, \ldots, N$; IF PAIR (i) = 0, then node i is unmatched.
NOTMAT	Number of unmatched nodes.

Working storages:

FWDARC	Integer vector of length M2; FWDARC(i) is the ending node of the ith edge in the forward star representation of the graph.
ARCFIR	Integer vector of length N1; ARCFIR(i) is the number of the first edge starting at node i in the forward star representation of the graph.
ANCES	Integer vector of length N; ANCES(i) is the grandfather of node i.
QUEUE	Integer vector of length N; the queue of nodes.

OUTREE Logical vector of length N; OUTREE(i) indicates whether node i is in the tree.

D. Test example

Find a maximum matching in the following graph.

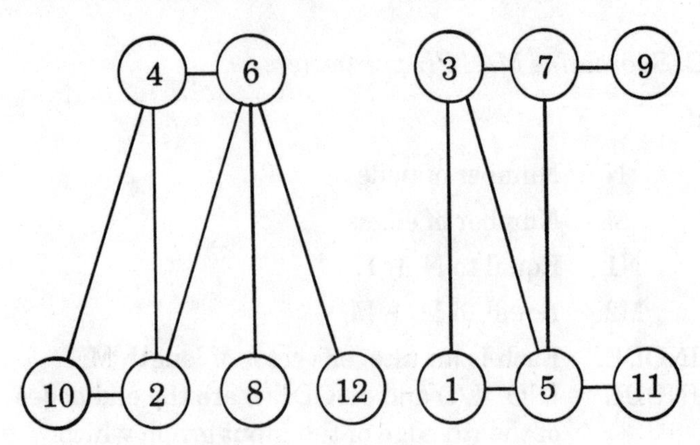

E. Program listing

MAIN PROGRAM

```
      INTEGER INODE(14),JNODE(14),PAIR(12),
     +        FWDARC(28),ARCFIR(13),
     +        ANCES(12),QUEUE(12)
      LOGICAL OUTREE(12)
      DATA INODE / 6,9,3, 4,11,6,4,5, 6,10,3,4,1,3/,
     +     JNODE / 2,7,7,10, 5,8,6,7,12, 2,1,2,5,5/
      N = 12
      M = 14
      N1 = N + 1
      M2 = M + M
      CALL MATCH(N,M,N1,M2,INODE,JNODE,
     +           PAIR,NOTMAT,
     +           FWDARC,ARCFIR,ANCES,
     +           QUEUE,OUTREE)
      WRITE(*,10) (PAIR(I),I=1,N)
   10 FORMAT(' ARRAY PAIR: ',12I3)
      WRITE(*,20) NOTMAT
```

```
   20    FORMAT(/' NUMBER OF UNMATCHED',
     +             ' NODES = ',I3)
         STOP
         END
```

OUTPUT RESULTS

```
ARRAY PAIR: 3  6  1  10  11  2  9  0  7  4  5  0
NUMBER OF UNMATCHED NODES = 2
```

```
         SUBROUTINE MATCH(N,M,N1,M2, INODE,
     +                        JNODE,PAIR,
     +                        NOTMAT,
     +                        FWDARC,ARCFIR,
     +                        ANCES,QUEUE,
     +                        OUTREE)
C
C        Maximum cardinality matching
C
         INTEGER INODE(M),JNODE(M),PAIR(N),
     +           FWDARC(M2),ARCFIR(N1),
     +           ANCES(N),QUEUE(N)
         LOGICAL OUTREE(N),NEWNOD,NOPATH
C
C        Set up the forward star representation of the
C          graph
C
         K = 0
         DO 20 I = 1, N
           ARCFIR(I) = K + 1
           DO 10 J = 1, M
             IF (INODE(J) .EQ. I) THEN
               K = K + 1
               FWDARC(K) = JNODE(J)
             ELSE
               IF (JNODE(J) .EQ. I) THEN
                 K = K + 1
```

```
                FWDARC(K) = INODE(J)
              ENDIF
            ENDIF
10        CONTINUE
20      CONTINUE
        ARCFIR(N1) = M + 1
C
C       All nodes are not matched initially
C
        NOTMAT = N
        DO 30 I = 1, N
30        PAIR(I) = 0
C
        DO 50 I = 1, N
          IF (PAIR(I) .EQ. 0) THEN
            J = ARCFIR(I)
            K = ARCFIR(I + 1) - 1
40          IF ((PAIR(FWDARC(J)) .NE. 0) .AND.
     +          (J .LT. K)) THEN
              J = J + 1
              GOTO 40
            ENDIF
            IF (PAIR(FWDARC(J)) .EQ. 0) THEN
C
C             Match a pair of nodes
C
              PAIR(FWDARC(J)) = I
              PAIR(I) = FWDARC(J)
              NOTMAT = NOTMAT - 2
            ENDIF
          ENDIF
50      CONTINUE
C
        DO 110 ISTART = 1, N
          IF ((NOTMAT .GE. 2) .AND. (PAIR(ISTART)
     +       .EQ. 0)) THEN
C
C           ISTART is not yet matched
C
```

```
      DO 60 I = 1, N
60       OUTREE(I) = .TRUE.
      OUTREE(ISTART) = .FALSE.
C
C     Put the root in the queue
C
      QUEUE(1) = ISTART
      IFIRST = 1
      ILAST = 1
      NOPATH = .TRUE.
70    NODEI = QUEUE(IFIRST)
      IFIRST = IFIRST + 1
      NODEU = ARCFIR(NODEI)
      NODEW = ARCFIR(NODEI + 1) - 1
80    IF (NOPATH .AND. (NODEU .LE.
     +   NODEW)) THEN
C
C        Examine the neighbor of NODEI
C
         IF (OUTREE(FWDARC(NODEU))) THEN
            NEIGH2 = FWDARC(NODEU)
            NODEJ = PAIR(NEIGH2)
            IF (NODEJ .EQ. 0) THEN
C
C              An augmentation path is found
C
               PAIR(NEIGH2) = NODEI
90                NEIGH1 = PAIR(NODEI)
                  PAIR(NODEI) = NEIGH2
                  IF (NEIGH1 .NE. 0) THEN
                     NODEI = ANCES(NODEI)
                     PAIR(NEIGH1) = NODEI
                     NEIGH2 = NEIGH1
                  ENDIF
               IF (NEIGH1 .NE. 0) GOTO 90
               NOTMAT = NOTMAT - 2
               NOPATH = .FALSE.
            ELSE
               IF (NODEJ .NE. NODEI) THEN
```

```
                IF (NODEI .EQ. ISTART) THEN
                  NEWNOD = .TRUE.
                ELSE
                  NODEV = ANCES(NODEI)
100               IF ((NODEV .NE. ISTART)
     +              .AND.
     +               (NODEV .NE. NEIGH2))
     +                  THEN
                    NODEV = ANCES(NODEV)
                    GOTO 100
                  ENDIF
                  IF (NODEV .EQ. ISTART) THEN
                    NEWNOD = .TRUE.
                  ELSE
                    NEWNOD = .FALSE.
                  ENDIF
                ENDIF
                IF (NEWNOD) THEN
C
C                 Add a tree edge
C
                  OUTREE(NEIGH2) = .FALSE.
                  ANCES(NODEJ) = NODEI
                  ILAST = ILAST + 1
                  QUEUE(ILAST) = NODEJ
                ENDIF
              ENDIF
            ENDIF
          ENDIF
          NODEU = NODEU + 1
          GOTO 80
        ENDIF
        IF (NOPATH .AND. (IFIRST .LE. ILAST))
     +    GOTO 70
      ENDIF
110   CONTINUE
C
      RETURN
      END
```

7

PLANARITY TESTING

A. Problem description

A graph is said to be *planar* if it can be drawn on a plane without any crossing edges. The problem is to decide whether a given undirected graph is planar.

B. Method

The basic strategy of the algorithm starts by finding a cycle C in the given graph G. Then, the graph $G - C$ is decomposed into edge-disjoint paths, and the paths are added to cycle C one at a time while keeping the embedding planar. If the embedding is successful, then G is planar; otherwise, it is nonplanar.

Let n be the number of nodes and m be the number of edges in the input graph G. The planarity testing algorithm can be described as follows.

STEP 1. If $m > 3n - 6$, then return the message "the input graph is nonplanar" and stop.

STEP 2. Perform a depth-first search on graph G to obtain a digraph D so that the edges of G are divided into tree edges and backward edges. During the search, compute the low point functions $L1$ and $L2$, where

> $L1(v)$ is the lowest node reachable from node v by a sequence of tree edges followed by at most one backward edge, and
>
> $L2(v)$ is the second lowest node reachable from node v in this manner.

STEP 3. Reorder the adjacency lists of D using a radix sort.

STEP 4. Use the low point functions computed from Step 2 and the ordering of edges from Step 3 to choose a particular adjacency structure so that the nodes of D are numbered in the order they are reached during any depth-first search for D without changing the adjacency structure.

STEP 5. Perform a second depth-first search to select edges in the reverse order to that given by the adjacency structure. The purpose of this search is to prepare for the path addition process in the next step.

STEP 6. Perform a third depth-first search to generate one cycle and a number of edge-disjoint paths. Each generated path is added to a planar embedding which contains the cycle and all the previously generated paths.

Note that any two paths may not constrain each other, or they may constrain each other to have either the same embedding or the opposite embedding. These dependency relations among paths can be viewed as a dependency graph H.

STEP 7. This last step makes use of the fact that dependency graph H is two-colorable if and only if the original graph G is planar Now, use a depth-first search to color graph H.

If H is two-colorable, then return the message "the input graph is planar"; otherwise, return the message "the input graph is nonplanar" and stop.

The computation of the method is bounded by $O(n)$.

C. Subroutine PLANE parameters

Input:

N	Number of nodes.
M	Number of edges.
INODE, JNODE	Each is an integer vector of length M; INODE(i) and JNODE(i) are the two end nodes of the ith edge in the undirected graph; the input graph is assumed to be connected.
N2	Equal to N + N.
M2	Equal to M + M.
M22	Equal to M + M + 2.
M33	Equal to M + M + M + 3.
NM2	Equal to N + M + M.
N2M	Equal to N + N + M.
NMP2	Equal to M − N + 2.
M7N5	Equal to 7 * M − 5 * N + 2

Output:

PLANAR	Boolean variable; PLANAR = TRUE if the graph is planar; PLANAR = FALSE if the graph is nonplanar.

Working storages:

MARK	Integer vector of length N; marking the nodes during a depth-first search.
TRAIL	Integer vector of length N; storing the path in the path-finding process.

SMALL1	Integer vector of length N; array for the first low point function.
SMALL2	Integer vector of length N; array for the second low point function.
DESCP	Integer vector of length N; number of descendants in the tree.
PINDEX	Integer vector of length N2; array for the highest start node in a path.
PARM1	Integer vector of length M; auxiliary array for the path-finding process.
PARM2	Integer vector of length M; auxiliary array for the path-finding process.
PAINT	Integer vector of length NMP2; coloring of nodes.
START	Integer vector of length NMP2; array for the starting node of a path in the path-finding process.
FINISH	Integer vector of length NMP2; array for the finishing node of a path in the path-finding process.
STACKE	Integer vector of length M2; stack for the edges during the search.
FIRST	Integer vector of length NM2; structural representation of the graph.
SECOND	Integer vector of length NM2; structural representation of the graph.
SORTN	Integer vector of length N2M; array for the sorting of the edges.
SORTP1	Integer vector of length N2M; pointer used in the sorting process.
SORTP2	Integer vector of length N2M; pointer used in the sorting process.
AUXPF1	Integer vector of length M22; auxiliary array used in the path-finding process.

AUXPF2	Integer vector of length M22; auxiliary array used in the path-finding process.
AUXPF3	Integer vector of length M33; auxiliary array used in the path-finding process.
AUXPF4	Integer vector of length M33; auxiliary array used in the path-finding process.
STAKC1	Integer vector of length M22; stack used in the coloring process.
STAKC2	Integer vector of length M22; stack used in the coloring process.
STAKC3	Integer vector of length M22; stack used in the coloring process.
STAKC4	Integer vector of length M22; stack used in the coloring process.
ARCSEC	Integer vector of length M7N5; array for the next edge in the coloring process.
ARCTOP	Integer vector of length M7N5; array for the first edge in the coloring process.
EXAMIN	Logical vector of length NMP2; array to mark whether the node has been examined in the coloring process.
ARCTYP	Logical vector of length M7N5; array to distinguish the edge type in the coloring process.

D. Test example

Consider the following graph of 10 nodes and 22 edges. Determine whether it is planar.

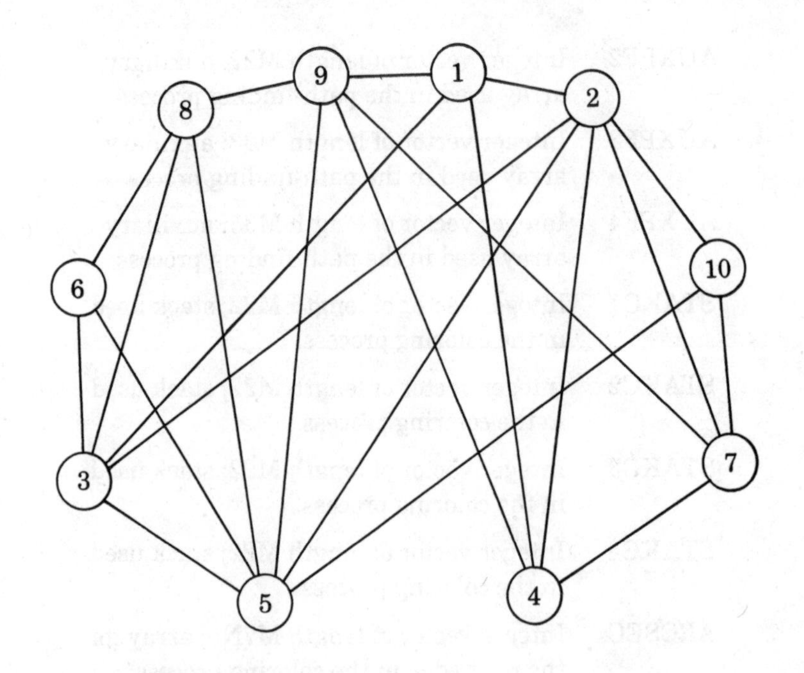

E. Program listing

MAIN PROGRAM

```
      IMPLICIT INTEGER (A-Z)
      INTEGER INODE(22),JNODE(22),MARK(10),
     +        TRAIL(10),SMALL1(10),
     +        SMALL2(10),DESCP(10),
     +        PINDEX(20),PARM1(22),
     +        PARM2(22),
     +        PAINT(14),START(14),FINISH(14),
     +        STACKE(44),FIRST(54),
     +        SECOND(54),SORTN(42),
     +        SORTP1(42),SORTP2(42),
     +        AUXPF1(46),
     +        AUXPF2(46),AUXPF3(69),
     +        AUXPF4(69),STAKC1(46),
     +        STAKC2(46),
     +        STAKC3(46),STAKC4(46),
     +        ARCSEC(106),ARCTOP(106)
```

```
      LOGICAL PLANAR,EXAMIN(14),
     +  ARCTYP(106)
      DATA INODE /9,3,10,2,6,7, 7,1,8,1,10,5,1,4,
     +                1,3,9,3,2,2,5,3/,
     +     JNODE /8,2, 5,4,5,9,10,3,6,2, 2,8,9,7,
     +                4,5,4,6,7,5,9,8/
C
      N = 10
      M = 22
      N2 = N + N
      M2 = M + M
      M22 = M + M + 2
      M33 = M + M + M + 3
      NM2 = N + M + M
      N2M = N + N + M
      NMP2 = M - N + 2
      M7N5 = 7 * M - 5 * N + 2
      CALL PLANE(N,M,INODE,JNODE,N2,M2,
     +           M22,M33,NM2,N2M,
     +           NMP2,M7N5,
     +           PLANAR,MARK,TRAIL,
     +           SMALL1,SMALL2,DESCP,
     +           PINDEX,PARM1,
     +           PARM2,PAINT,START,
     +           FINISH,STACKE,FIRST,
     +           SECOND,SORTN,
     +           SORTP1,SORTP2,AUXPF1,
     +           AUXPF2,AUXPF3,AUXPF4,
     +           STAKC1,
     +           STAKC2,STAKC3,STAKC4,
     +           ARCSEC,ARCTOP,EXAMIN,
     +           ARCTYP)
      IF (PLANAR) THEN
        WRITE(*,10)
10      FORMAT(/' THE INPUT GRAPH IS',
     +         ' PLANAR')
      ELSE
        WRITE(*,20)
20      FORMAT(/' THE INPUT GRAPH IS',
```

```
     +               ' NONPLANAR')
      ENDIF
      STOP
      END
```

OUTPUT RESULTS

```
  THE INPUT GRAPH IS PLANAR
```

```
      SUBROUTINE PLANE (N,M,INODE,JNODE,
     +                  N2,M2,M22,M33,
     +                  NM2,N2M,NMP2,
     +                  M7N5,PLANAR,
     +                  MARK,TRAIL,
     +                  SMALL1,SMALL2,
     +                  DESCP,
     +                  PINDEX,PARM1,
     +                  PARM2,PAINT,
     +                  START,FINISH,
     +                  STACKE,
     +                  FIRST,SECOND,
     +                  SORTN,SORTP1,
     +                  SORTP2,AUXPF1,
     +                  AUXPF2,AUXPF3,
     +                  AUXPF4,STAKC1,
     +                  STAKC2,STAKC3,
     +                  STAKC4,ARCSEC,
     +                  ARCTOP,EXAMIN,
     +                  ARCTYP)
C
C     Planarity graph-testing
C
      IMPLICIT INTEGER (A-Z)
      INTEGER INODE(M),JNODE(M),MARK(N),
     +        TRAIL(N),SMALL1(N),SMALL2(N),
     +        DESCP(N),PINDEX(N2),PARM1(M),
     +        PARM2(M),PAINT(NMP2),
     +        START(NMP2),FINISH(NMP2),
     +        STACKE(M2),FIRST(NM2),
```

```
     +              SECOND(NM2),SORTN(N2M),
     +              SORTP1(N2M),SORTP2(N2M),
     +              AUXPF1(M22),AUXPF2(M22),
     +              AUXPF3(M33),AUXPF4(M33),
     +              STAKC1(M22),STAKC2(M22),
     +              STAKC3(M22),STAKC4(M22),
     +              ARCSEC(M7N5),ARCTOP(M7N5)
      LOGICAL PLANAR,EXAMIN(NMP2),
     +  ARCTYP(M7N5),MIDDLE,FAIL
C
      IF (M .GT. 3*N-6) THEN
        PLANAR = .FALSE.
        RETURN
      ENDIF
C
C     Set up the graph representation
C
      DO 10 I = 1, N
  10    SECOND(I) = 0
      MTOTAL = N
      DO 20 I = 1, M
        NODE1 = INODE(I)
        NODE2 = JNODE(I)
        MTOTAL = MTOTAL + 1
        SECOND(MTOTAL) = SECOND(NODE1)
        SECOND(NODE1) = MTOTAL
        FIRST(MTOTAL) = NODE2
        MTOTAL = MTOTAL + 1
        SECOND(MTOTAL) = SECOND(NODE2)
        SECOND(NODE2) = MTOTAL
        FIRST(MTOTAL) = NODE1
  20  CONTINUE
C
C     Initial depth-first search; compute the low point
C       functions.
C
      DO 30 I = 1, N
        MARK(I) = 0
        SMALL1(I) = N + 1
```

```
30       SMALL2(I) = N + 1
      SNUM = 1
      PW12 = 0
      MARK(1) = 1
      PARM1(1) = 1
      PARM2(1) = 0
      LEVEL = 1
      MIDDLE = .FALSE.
40    CALL DFS1(N,M,M2,NM2,LEVEL,MIDDLE,
     +           SNUM,PW12,MARK,
     +           SMALL1,SMALL2,PARM1,
     +           PARM2,STACKE,FIRST,
     +           SECOND)
      IF (LEVEL .GT. 1) GOTO 40
C
      DO 50 I = 1, N
        IF (SMALL2(I) .GE. MARK(I))
     +    SMALL2(I) = SMALL1(I)
50    CONTINUE
C
C     Radix sort
C
      MTOTAL = N2
      K = N2
      DO 60 I = 1, N2
60      SORTN(I) = 0
      DO 70 I = 2, M2, 2
        K = K + 1
        SORTP1(K) = STACKE(I-1)
        TNODE = STACKE(I)
        SORTP2(K) = TNODE
        IF (MARK(TNODE) .LT. MARK(SORTP1(K)))
     +    THEN
          J = 2 * MARK(TNODE) - 1
          SORTN(K) = SORTN(J)
          SORTN(J) = K
        ELSE
          IF (SMALL2(TNODE) .GE.
     +      MARK(SORTP1(K))) THEN
```

```
           J = 2 * SMALL1(TNODE) - 1
           SORTN(K) = SORTN(J)
           SORTN(J) = K
         ELSE
           J = 2 * SMALL1(TNODE)
           SORTN(K) = SORTN(J)
           SORTN(J) = K
         ENDIF
       ENDIF
 70  CONTINUE
C
     DO 90 I = 1, N2
       J = SORTN(I)
 80    IF (J .NE. 0) THEN
         NODE1 = SORTP1(J)
         NODE2 = SORTP2(J)
         MTOTAL = MTOTAL + 1
         SECOND(MTOTAL) = SECOND(NODE1)
         SECOND(NODE1) = MTOTAL
         FIRST(MTOTAL) = NODE2
         J = SORTN(J)
         GOTO 80
       ENDIF
 90  CONTINUE
C
C    Second depth-first search
C
     DO 100 I = 2, N
100    MARK(I) = 0
     PW12 = 0
     SNUM = 1
     TRAIL(1) = 1
     PARM1(1) = 1
     START(1) = 0
     FINISH(1) = 0
     LEVEL = 1
     MIDDLE = .FALSE.
C
110  CALL DFS2(N,M,M2,NM2,LEVEL,MIDDLE,
```

```
     +             SNUM,
     +             PW12,MARK,PARM1,STACKE,
     +             FIRST,SECOND)
C
      IF (LEVEL .GT. 1) GOTO 110
C
      MTOTAL = N
      DO 120 I = 1, M
        J = I + I
        NODE1 = STACKE(J-1)
        NODE2 = STACKE(J)
        MTOTAL = MTOTAL + 1
        SECOND(MTOTAL) = SECOND(NODE1)
        SECOND(NODE1) = MTOTAL
        FIRST(MTOTAL) = NODE2
 120  CONTINUE
C
C     Path decomposition; construction of the
C       dependency graph.
C
      PW18 = 0
      PW19 = 0
      PW24 = 0
      PW25 = 0
      INITP = 0
      PNUM = 1
      AUXPF1(1) = 0
      AUXPF1(2) = 0
      AUXPF2(1) = 0
      AUXPF2(2) = 0
      AUXPF3(1) = 0
      AUXPF3(2) = N + 1
      AUXPF3(3) = 0
      AUXPF4(1) = 0
      AUXPF4(2) = N + 1
      AUXPF4(3) = 0
      DO 130 I = 1, N2
 130    PINDEX(I) = 0
      NEXTE = M - N + 1
```

```
      DO 140 I = 1, M7N5
140     ARCSEC(I) = 0
      SNODE = N
      DESCP(1) = N
      PARM1(1) = 1
      LEVEL = 1
      MIDDLE = .FALSE.
C
150   CALL DECOMP(N,M,N2,M22,M33,NM2,
     +             NMP2,M7N5,LEVEL,
     +             MIDDLE,INITP,
     +             SNODE,PNUM,NEXTE,
     +             PW18,PW19,PW24,PW25,
     +             TRAIL,DESCP,
     +             PINDEX,PARM1,START,
     +             FINISH,FIRST,SECOND,
     +             AUXPF1,
     +             AUXPF2,AUXPF3,AUXPF4,
     +             ARCSEC,ARCTOP,ARCTYP)
      IF (LEVEL .GT. 1) GOTO 150
C
C     Perform the two-coloring
C
      PNUM = PNUM - 1
      DO 160 I = 1, NMP2
160     PAINT(I) = 0
      J = PNUM + 1
      DO 170 I = 2, J
170     EXAMIN(I) = .TRUE.
      TNUM = 1
180   IF (TNUM .LE. PNUM) THEN
        PARM1(1) = TNUM
        PAINT(TNUM) = 1
        EXAMIN(TNUM) = .FALSE.
        LEVEL = 1
        MIDDLE = .FALSE.
190     CALL COLOR2(M,NMP2,M7N5,LEVEL,
     +               MIDDLE,FAIL,PARM1,
     +               PAINT,ARCSEC,ARCTOP,
```

```
     +                     EXAMIN,ARCTYP)
          IF (FAIL) THEN
            PLANAR = .FALSE.
            RETURN
          ENDIF
          IF (LEVEL .GT. 1) GOTO 190
200       IF (.NOT. EXAMIN(TNUM)) THEN
            TNUM = TNUM + 1
            GOTO 200
          ENDIF
          GOTO 180
        ENDIF
C
        PW20 = 0
        PW21 = 0
        PW22 = 0
        PW23 = 0
        STAKC1(1) = 0
        STAKC1(2) = 0
        STAKC2(1) = 0
        STAKC2(2) = 0
        STAKC3(1) = 0
        STAKC3(2) = 0
        STAKC4(1) = 0
        STAKC4(2) = 0
C
        DO 250 I = 1, PNUM
          QNODE = START(I+1)
          TNODE = FINISH(I+1)
210       IF (QNODE .LE. STAKC1(PW20+2)) THEN
            PW20 = PW20 - 2
            GOTO 210
          ENDIF
220       IF (QNODE .LE. STAKC2(PW21+2)) THEN
            PW21 = PW21 - 2
            GOTO 220
          ENDIF
```

```
230       IF (QNODE .LE. STAKC3(PW22+2)) THEN
            PW22 = PW22 - 2
            GOTO 230
          ENDIF
240       IF (QNODE .LE. STAKC4(PW23+2)) THEN
            PW23 = PW23 - 2
            GOTO 240
          ENDIF
          IF (PAINT(I) .EQ. 1) THEN
            IF (FINISH(TRAIL(QNODE)+1) .NE.
    +         TNODE) THEN
              IF (TNODE .LT. STAKC2(PW21+2))
    +           THEN
                PLANAR = .FALSE.
                RETURN
              ENDIF
              IF (TNODE .LT. STAKC3(PW22+2))
    +           THEN
                PLANAR = .FALSE.
                RETURN
              ENDIF
              PW22 + PW22 + 2
              STAKC3(PW22 + 1) = I
              STAKC3(PW22 + 2) = TNODE
            ELSE
              IF ((TNODE .LT. STAKC3(PW22+2))
    +           .AND.
    +           (START(STAKC3(PW22+1)+1) .LE.
    +             DESCP(QNODE))) THEN
                PLANAR = .FALSE.
                RETURN
              ENDIF
              PW20 = PW20 + 2
              STAKC1(PW20 + 1) = I
              STAKC1(PW20 + 2) = QNODE
            ENDIF
          ELSE
```

```
          IF (FINISH(TRAIL(QNODE)+1) .NE.
     +       TNODE) THEN
             IF (TNODE .LT. STAKC1(PW20+2))
     +         THEN
               PLANAR = .FALSE.
               RETURN
             ENDIF
             IF (TNODE .LT. STAKC4(PW23+2))
     +         THEN
               PLANAR = .FALSE.
               RETURN
             ENDIF
             PW23 = PW23 + 2
             STAKC4(PW23 + 1) = I
             STAKC4(PW23 + 2) = TNODE
           ELSE
             IF ((TNODE .LT. STAKC4(PW23+2))
     +         .AND.
     +         (START(STAKC4(PW23+1)+1) .LE.
     +           DESCP(QNODE))) THEN
               PLANAR = .FALSE.
               RETURN
             ENDIF
             PW21 = PW21 + 2
             STAKC2(PW21 + 1) = I
             STAKC2(PW21 + 2) = QNODE
           ENDIF
         ENDIF
 250   CONTINUE
       PLANAR = .TRUE.
C
       RETURN
       END


       SUBROUTINE DFS1(N,M,M2,NM2,LEVEL,
     +                     MIDDLE,SNUM,
     +                     PW12,MARK,
     +                     SMALL1,SMALL2,
```

```
     +                  PARM1,PARM2,
     +                  STACKE,FIRST,
     +                  SECOND)
C
C    Initial depth-first search; compute the low
C      QNODE functions (this subprogram is used
C      by subroutine PLANE).
C
      IMPLICIT INTEGER (A-Z)
      INTEGER MARK(N),SMALL1(N),SMALL2(N),
     +            PARM1(M),PARM2(M),
     +            STACKE(M2),FIRST(NM2),
     +            SECOND(NM2)
      LOGICAL MIDDLE
C
      IF (MIDDLE) GOTO 20
      QNODE = PARM1(LEVEL)
      PNODE = PARM2(LEVEL)
 10   IF (SECOND(QNODE) .GT. 0) THEN
        TNODE = FIRST(SECOND(QNODE))
        SECOND(QNODE)
     +     = SECOND(SECOND(QNODE))
        IF ((MARK(TNODE) .LT. MARK(QNODE))
     +    .AND. (TNODE .NE. PNODE)) THEN
          PW12 = PW12 + 2
          STACKE(PW12-1) = QNODE
          STACKE(PW12) = TNODE
          IF (MARK(TNODE) .EQ. 0) THEN
            SNUM = SNUM + 1
            MARK(TNODE) = SNUM
            LEVEL = LEVEL + 1
            PARM1(LEVEL) = TNODE
            PARM2(LEVEL) = QNODE
            MIDDLE = .FALSE.
            RETURN
C
 20         TNODE = PARM1(LEVEL)
            QNODE = PARM2(LEVEL)
            LEVEL = LEVEL - 1
```

```
      PNODE = PARM2(LEVEL)
      IF (SMALL1(TNODE) .LT.
+       SMALL1(QNODE)) THEN
        ITEMP1 = SMALL2(TNODE)
        ITEMP2 = SMALL1(QNODE)
        IF (ITEMP1 .LT. ITEMPT2) THEN
          SMALL2(QNODE) = ITEMP1
        ELSE
          SMALL2(QNODE) = ITEMP2
        ENDIF
        SMALL1(QNODE)
+         = SMALL1(TNODE)
      ELSE
        IF (SMALL1(TNODE) .EQ.
+         SMALL1(QNODE)) THEN
          ITEMP1 = SMALL2(TNODE)
          ITEMP2 = SMALL2(QNODE)
          IF (ITEMP1 .LT. ITEMP2) THEN
            SMALL2(QNODE) = ITEMP1
          ELSE
            SMALL2(QNODE) = ITEMP2
          ENDIF
        ELSE
          ITEMP1 = SMALL1(TNODE)
          ITEMP2 = SMALL2(QNODE)
          IF (ITEMP1 .LT. ITEMP2) THEN
            SMALL2(QNODE) = ITEMP1
          ELSE
            SMALL2(QNODE) = ITEMP2
          ENDIF
        ENDIF
      ENDIF
    ELSE
      IF (MARK(TNODE) .LT.
+       SMALL1(QNODE)) THEN
        SMALL2(QNODE)
+         = SMALL1(QNODE)
        SMALL1(QNODE) = MARK(TNODE)
      ELSE
```

```
            IF (MARK(TNODE) .GT.
     +         SMALL1(QNODE)) THEN
               ITEMP1 = MARK(TNODE)
               ITEMP2 = SMALL2(QNODE)
               IF (ITEMP1 .LT. ITEMP2) THEN
                 SMALL2(QNODE) = ITEMP1
               ELSE
                 SMALL2(QNODE) = ITEMP2
               ENDIF
            ENDIF
          ENDIF
        ENDIF
      ENDIF
      GOTO 10
    ENDIF
    MIDDLE = .TRUE.
C
    RETURN
    END


    SUBROUTINE DFS2(N,M,M2,NM2,LEVEL,
   +                    MIDDLE,SNUM,
   +                    PW12,MARK,PARM1,
   +                    STACKE,FIRST,
   +                    SECOND)
C
C   Second depth-first search (this subprogram is
C     used by subroutine PLANE)
C
    IMPLICIT INTEGER (A-Z)
    INTEGER MARK(N),PARM1(M),STACK(M2),
   +          FIRST(NM2),SECOND(NM2)
    LOGICAL MIDDLE
C
    IF (MIDDLE) THEN
      TNODE = PARM1(LEVEL)
      LEVEL = LEVEL - 1
      QNODE = PARM1(LEVEL)
```

```
        PW12 = PW12 + 2
        STACKE(PW12-1) = MARK(QNODE)
        STACKE(PW12) = MARK(TNODE)
      ELSE
        QNODE = PARM1(LEVEL)
      ENDIF
C
 10   IF (SECOND(QNODE) .GT. 0) THEN
        TNODE = FIRST(SECOND(QNODE))
        SECOND(QNODE)
     +      = SECOND(SECOND(QNODE))
        IF (MARK(TNODE) .EQ. 0) THEN
          SNUM = SNUM + 1
          MARK(TNODE) = SNUM
          LEVEL = LEVEL + 1
          PARM1(LEVEL) = TNODE
          MIDDLE = .FALSE.
          RETURN
        ENDIF
        PW12 = PW12 + 2
        STACKE(PW12-1) = MARK(QNODE)
        STACKE(PW12) = MARK(TNODE)
        GOTO 10
      ENDIF
      MIDDLE = .TRUE.
C
      RETURN
      END


      SUBROUTINE DECOMP(N,M,N2,M22,M33,
     +                  NM2,NMP2,M7N5,
     +                  LEVEL,MIDDLE,
     +                  INITP,SNODE,
     +                  PNUM,NEXTE,
     +                  PW18,PW19,PW24,
     +                  PW25,
     +                  TRAIL,DESCP,
     +                  PINDEX,PARM1,
```

```
     +                        START,FINISH,
     +                        FIRST,SECOND,
     +                        AUXPF1,AUXPF2,
     +                        AUXPF3,AUXPF4,
     +                        ARCSEC,ARCTOP,
     +                        ARCTYP)
C
C     Path decomposition (this subprogram is used
C       by subroutine PLANE)
C
      IMPLICIT INTEGER (A-Z)
      INTEGER TRAIL(N),DESCP(N),PINDEX(N2),
     +          PARM1(M),START(NMP2),
     +          FINISH(NMP2),FIRST(NM2),
     +          SECOND(NM2),AUXPF1(M22),
     +          AUXPF2(M22),AUXPF3(M33),
     +          AUXPF4(M33),ARCSEC(M7N5),
     +          ARCTOP(M7N5)
      LOGICAL MIDDLE,IND,ARCTYPE(M7N5)
C
      IF (MIDDLE) GOTO 20
      QNODE = PARM1(LEVEL)
   10 IF (SECOND(QNODE) .NE. 0) THEN
        TNODE = FIRST(SECOND(QNODE))
        SECOND(QNODE)
     +    = SECOND(SECOND(QNODE))
        IF (INITP .EQ. 0) INITP = QNODE
        IF (TNODE .GT. QNODE) THEN
          DESCP(TNODE) = SNODE
          TRAIL(TNODE) = PNUM
          LEVEL = LEVEL + 1
          PARM1(LEVEL) = TNODE
          MIDDLE = .FALSE.
          RETURN
C
   20     TNODE = PARM1(LEVEL)
          LEVEL = LEVEL – 1
          QNODE = PARM1(LEVEL)
          SNODE = TNODE – 1
```

```
      INITP = 0
30    IF (QNODE .LE. AUXPF2(PW19 + 2))
   +    THEN
        PW19 = PW19 - 2
        GOTO 30
      ENDIF
C
40    IF (QNODE .LE. AUXPF1(PW18 + 2))
   +    THEN
        PW18 = PW18 - 2
        GOTO 40
      ENDIF
C
50    IF (QNODE .LE. AUXPF3(PW24 + 3))
   +    THEN
        PW24 = PW24 - 3
        GOTO 50
      ENDIF
C
60    IF (QNODE .LE. AUXPF4(PW25 + 3))
   +    THEN
        PW25 = PW25 - 3
        GOTO 60
      ENDIF
      IND = .FALSE.
      QNODE2 = QNODE + QNODE
70    IF ((PINDEX(QNODE2 - 1)
   +    .GT. AUXPF3(PW24 + 2)) .AND.
   +     (QNODE .LT. AUXPF3(PW24 + 2))
   +        .AND.
   +      (PINDEX(QNODE2) .LT.
   +        AUXPF3(PW24 + 1))) THEN
        IND = .TRUE.
        NODE1 = PINDEX(QNODE2)
        NODE2 = AUXPF3(PW24 + 1)
        NEXTE = NEXTE + 1
        ARCSEC(NEXTE) = ARCSEC(NODE1)
        ARCSEC(NODE1) = NEXTE
        ARCTOP(NEXTE) = NODE2
```

```
            NODE1 = AUXPF3(PW24 + 1)
            NODE2 = PINDEX(QNODE2)
            NEXTE = NEXTE + 1
            ARCSEC(NEXTE) = ARCSEC(NODE1)
            ARCSEC(NODE1) = NEXTE
            ARCTOP(NEXTE) = NODE2
            ARCTYP(NEXTE - 1) = .FALSE.
            ARCTYP(NEXTE) = .FALSE.
            PW24 = PW24 - 3
            GOTO 70
          ENDIF
          IF (IND) PW24 = PW24 + 3
          PINDEX(QNODE2 - 1) = 0
          PINDEX(QNODE2) = 0
        ELSE
          START(PNUM + 1) = INITP
          FINISH(PNUM + 1) = TNODE
          IND = .FALSE.
          IF (AUXPF1(PW18+2) .NE. 0) THEN
            PW19 = PW19 + 2
            AUXPF2(PW19+1)
     +        = AUXPF1(PW18+1)
            AUXPF2(PW19+2)
     +        =AUXPF1(PW18+2)
          ENDIF
          IF (FINISH(AUXPF1(PW18+1) + 1)
     +      .NE. TNODE) THEN
80          IF (TNODE .LT. AUXPF2(PW19+2))
     +        THEN
              NODE1 = PNUM
              NODE2 = AUXPF2(PW19 + 1)
              NEXTE = NEXTE + 1
              ARCSEC(NEXTE)
     +          = ARCSEC(NODE1)
              ARCSEC(NODE1) = NEXTE
              ARCTOP(NEXTE) = NODE2
              NODE1 = AUXPF2(PW19 + 1)
              NODE2 = PNUM
              NEXTE = NEXTE + 1
```

```
              ARCSEC(NEXTE)
     +          = ARCSEC(NODE1)
              ARCSEC(NODE1) = NEXTE
              ARCTOP(NEXTE) = NODE2
              ARCTYP(NEXTE - 1) = .TRUE.
              ARCTYP(NEXTE) = .TRUE.
              IND = .TRUE.
              PW19 = PW19 - 2
              GOTO 80
            ENDIF
            IF (IND) PW19 = PW19 + 2
            IND = .FALSE.
90          IF ((TNODE .LT. AUXPF3(PW24+3))
     +        .AND.
     +          (INITP .LT. AUXPF3(PW24+2)))
     +            THEN
              NODE1 = PNUM
              NODE2 = AUXPF3(PW24 + 1)
              NEXTE = NEXTE + 1
              ARCSEC(NEXTE)
     +          = ARCSEC(NODE1)
              ARCSEC(NODE1) = NEXTE
              ARCTOP(NEXTE) = NODE2
              NODE1 = AUXPF3(PW24 + 1)
              NODE2 = PNUM
              NEXTE = NEXTE + 1
              ARCSEC(NEXTE)
     +          = ARCSEC(NODE1)
              ARCSEC(NODE1) = NEXTE
              ARCTOP(NEXTE) = NODE2
              ARCTYP(NEXTE - 1) = .FALSE.
              ARCTYP(NEXTE) = .FALSE.
              PW24 = PW24 - 3
              GOTO 90
            ENDIF
100         IF ((TNODE .LT. AUXPF4(PW25+3))
     +        .AND.
     +          (INITP .LT. AUXPF4(PW25+2)))
     +            THEN
```

```
            PW25 = PW25 - 3
            GOTO 100
          ENDIF
          TNODE2 = TNODE + TNODE
          IF (INITP .GT. PINDEX(TNODE2-1))
     +      THEN
            PINDEX(TNODE2-1) = INITP
            PINDEX(TNODE2) = PNUM
          ENDIF
          PW24 = PW24 + 3
          AUXPF3(PW24+1) = PNUM
          AUXPF3(PW24+2) = INITP
          AUXPF3(PW24+3) = TNODE
          PW24 = PW25 + 3
          AUXPF4(PW25+1) = PNUM
          AUXPF4(PW25+2) = INITP
          AUXPF4(PW25+3) = TNODE
        ELSE
110       IF  ((TNODE  .LT.  AUXPF4(PW25+3))
     +      .AND.
     +       (INITP .LT. AUXPF4(PW25+2))
     +          .AND.
     +       (AUXPF4(PW25+2) .LE.
     +          DESCP(INITP))) THEN
            IND = .TRUE.
            NODE1 = PNUM
            NODE2 = AUXPF4(PW25 + 1)
            NEXTE = NEXTE + 1
            ARCSEC(NEXTE)
     +        = ARCSEC(NODE1)
            ARCSEC(NODE1) = NEXTE
            ARCTOP(NEXTE) = NODE2
            NODE1 = AUXPF4(PW25 + 1)
            NODE2 = PNUM
            NEXTE = NEXTE + 1
            ARCSEC(NEXTE)
     +        = ARCSEC(NODE1)
            ARCSEC(NODE1) = NEXTE
            ARCTOP(NEXTE) = NODE2
```

```
              ARCTYP(NEXTE - 1) = .FALSE.
              ARCTYP(NEXTE) = .FALSE.
              PW25 = PW25 - 3
              GOTO 110
            ENDIF
            IF (IND) PW25 = PW25 + 3
          ENDIF
          IF (QNODE .NE. INITP) THEN
            PW18 = PW18 + 2
            AUXPF1(PW18+1) = PNUM
            AUXPF1(PW18+2) = INITP
          ENDIF
          PNUM = PNUM + 1
          INITP = 0
        ENDIF
        GOTO 10
      ENDIF
      MIDDLE = .TRUE.
C
      RETURN
      END


      SUBROUTINE COLOR2(M,NMP2,M7N5,
     +                  LEVEL,MIDDLE,
     +                  FAIL,
     +                  PARM1,PAINT,
     +                  ARCSEC,ARCTOP,
     +                  EXAMIN,ARCTYP)
C
C     Two-coloring (this subprogram is used by
C       subroutine (PLANE)
C
      IMPLICIT INTEGER (A-Z)
      INTEGER PARM1(M),PAINT(NMP2),
     +           ARCSEC(M7N5),ARCTOP(M7N5)
      LOGICAL MIDDLE,EXAMIN(NMP2),
```

```
     +  ARCTYPE(M7N5),FAIL,DUM1,DUM2
C
        FAIL = .FALSE.
        IF (MIDDLE) THEN
          LEVEL = LEVEL – 1
          QNODE = PARM1(LEVEL)
        ELSE
          QNODE = PARM1(LEVEL)
        ENDIF
C
   10   IF (ARCSEC(QNODE) .NE. 0) THEN
          LINK = ARCSEC(QNODE)
          TNODE = ARCTOP(LINK)
          ARCSEC(QNODE) = ARCSEC(LINK)
          IF (PAINT(TNODE) .EQ. 0) THEN
            IF (ARCTYP(LINK)) THEN
              PAINT(TNODE) = PAINT(QNODE)
            ELSE
              PAINT(TNODE) = 3 – PAINT(QNODE)
            ENDIF
          ELSE
            DUM1 = PAINT(TNODE) .EQ.
     +        PAINT(QNODE)
            DUM2 = .NOT. ARCTYP(LINK)
            IF ((DUM1 .AND. DUM2) .OR.
     +          (.NOT.DUM1 .AND. .NOT.DUM2))
     +            THEN
              FAIL = .TRUE.
              RETURN
            ENDIF
          ENDIF
          IF (EXAMIN(TNODE)) THEN
            EXAMIN(TNODE) = .FALSE.
            LEVEL = LEVEL + 1
            PARM1(LEVEL) = TNODE
            MIDDLE = .FALSE.
            RETURN
```

```
            ENDIF
C
            GOTO 10
         ENDIF
         MIDDLE = .TRUE.
C
         RETURN
         END
```

APPENDIX I

GRAPH TERMINOLOGY

A *graph* consists of a set of *nodes* and a set of *edges*. Each edge e is a pair of distinct nodes which are called *end nodes* of edge e. A graph is *complete* if there is an edge between every pair of distinct nodes. An edge is said to be *directed* if its pair of end nodes is an ordered pair. A *directed graph*, or *digraph*, is a graph in which all edges are directed; otherwise, the graph is said to be *undirected*. If $e = (i, j)$ is an edge, then i and j are *adjacent nodes*. Node i and edge e are *incident* to each other, and so are j and e. If two edges e and f are incident with a common node, then e and f are *adjacent edges*. The *degree* of a node is the number of edges incident to the node. The *complement* H of a graph G is the graph with the same set of nodes, and two nodes are adjacent in H if and only if they are not adjacent in G.

A *walk* is a sequence of nodes in which two consecutive nodes are the end nodes of an edge in the graph. A *path* is a walk in which all nodes are distinct. A *directed path* is a path in a digraph. A *cycle* is a walk in which the first and last nodes are the same and all nodes are distinct. A graph is *connected* if there exists a path between every pair of nodes; otherwise, the graph is *disconnected.* A *strongly-connected* graph is a digraph in which there exists a directed path between every two nodes. A *subgraph* S of a graph G has all its nodes and edges in G. A maximal connected subgraph of G is called a *component* of G. A *spanning subgraph* is a subgraph containing all the nodes of G.

A *tree* is a connected graph with no cycles. A graph without cycles is a *forest.* A *spanning tree* of a graph G is a spanning subgraph of G that is a tree. A *spanning forest* of a graph G is a spanning subgraph of G that is a forest. A *binary tree* is a tree in which no node has a degree greater than two.

A *cut node* of a graph G is a node whose removal will increase the number of components of G. Similarly, a *bridge* is an edge whose removal will increase the number of components of G. A connected graph with no cut nodes is called a *block.*

The *node connectivity* of a graph is the minimum number of nodes whose removal will result in a disconnected graph or a graph with a single node. Similarly, the *edge connectivity* of a graph is the minimum number of edges whose removal will result in a disconnected graph or a graph with a single node. A graph is *k-connected* if the node connectivity of G is at least k.

APPENDIX II

LIST OF SUBROUTINES

BIBLIOGRAPHIC NOTES

1-1. The construction of a graph of largest possible connectivity with a given number of nodes and edges is given by F. Harary, "The Maximum Connectivity of a Graph," *Proceedings of the National Academy of Sciences of the United States of America*, Vol. 48, 1962, 1142-1146.

1-2. The application of network flow technique to find the edge connectivity of a graph is due to S. Even and R.E. Tarjan, "Network Flow and Testing Graph Connectivity," *SIAM Journal on Computing*, Vol. 4, 1975, 507-518. The maximum network flow algorithm is due to A.V. Karzanov, "Determining the Maximal Flow in a Network by the Method of Preflows," *Soviet Mathematics. Dolady.*, Vol. 15, 1974, 434-437.

1-3. The fundamental set of cycles are generated by an algorithm given by K. Paton, "An Algorithm for Finding a Fundamental Set of Cycles in a Graph," *Communications of the ACM*, Vol. 12, 1969, 514-518.

1-4. The paper by K. Paton, "An Algorithm for the Blocks and Cutnodes of a Graph," *Communications of the ACM*, Vol. 14, 1971, 468-475, shows how to enumerate the cut nodes and bridges of a graph by extending the fundamental cycle generation algorithm.

1-5. The application of a depth-first search to find the strongly-connected components of a digraph originates from R. Tarjan, "Depth-First Search and Linear Graph Algorithms," *SIAM Journal on Computing*, Vol. 1, 1972, 146-160.

1-6. The branch and bound algorithm for finding a minimal equivalent graph is given by S. Martello, "An Algorithm for Finding a Minimal Equivalent Graph of a Strongly-Connected Digraph," *Joint National ORSA/TIMS Meeting*, November 1977, Atlanta, Georgia.

1-7. The systematic enumerative method for finding all maximal independent sets in a graph is given by C. Bron and J. Kerbosch, "Algorithm 457—Finding All Cliques of an Undirected Graph," *Communications of the ACM*, Vol. 16, 1973, 575-577.

2-1. In this section, the algorithm for finding a shortest path from a specified source to a specified sink is due to T.A.J. Nicholson, "Finding the Shortest Route Between Two Points in a Network," *The Computer Journal*, Vol. 9, November 1966, 275-280.

Due to its worst-case characteristics, this algorithm has been generally ignored. However, in the thesis by H.K. DeWitt, "The Theory of Random Graphs With Applications to the Probabilistic Analysis of Optimization Algorithms," Ph.D. dissertation, Computer Science Department, University of California, Los Angeles, 1977, it is shown that the expected running time of Nicholson's algorithm on random graphs is only $O(n^{1/2}\log n)$, requiring far less computation than other known algorithms.

2-2. The efficient computer programming technique for finding all shortest-path lengths from a fixed node was first suggested by J.Y. Yen, "Finding the Lengths of All Shortest Paths in n-Node Nonnegative-Distance Complete Networks Using $n^3/2$ additions and n^3 comparisons," *Journal of the ACM*, Vol. 19, July 1972, 423-424.

2-3. The basic label-correcting method for finding a shortest path tree has been ascribed to Ford and Moore. L.R. Ford, Jr., "Network Flow Theory," The Rand Corporation, Report P-923, August 1956. E.F. Moore, "The Shortest Path Through a Maze," Proceedings of an international symposium on the theory of switching, Part II, April 2-5, 1957, *The Annals of the Computation Laboratory of Harvard University*, Vol. 30, Harvard University Press, Cambridge, Massachusetts, 1959, 285-292.

2-4. The all-pairs shortest-path algorithm was originally published by R.W. Floyd, "Algorithms 97—Shortest Path," *Communications of the ACM*, Vol. 5, 1962, 345.

2-5. The double-sweep algorithm for computing *k* shortest paths is due to D.R. Shier, "Iterative Methods for Determining the k Shortest Paths in a Network," *Networks*, Vol. 6, 1976, 205-229.

2-6. The method of finding the *k* shortest paths without repeated nodes is given by J. Y. Yen, "Finding the k Shortest Loopless Paths in a Network," *Management Science*, Vol. 17, 1971, 712-716.

3-1. Fleury gives a good algorithm for constructing an Euler circuit in an Eulerian graph; see E. Lucas, *Récréations Mathématiques* IV, Paris, 1921.

3-2. The backtrack method is commonly used in exhaustive search problems. As an example, see R.J. Walker, "An Enumerative Technique for a Class of Combinatorial Problems," *Proceedings of Symposia in Applied Mathematics*, American Mathematical Society, Providence, Rhode Island, Vol. 10, 1960, 91-94.

4-1. The backtracking algorithm in coloring a graph was first proposed by J.R. Brown, "Chromatic Scheduling and the Chromatic Number Problem," *Management Science*, Vol. 19, December, Part I, 1972, 456-463.

4-2. An exposition on chromatic polynomials is given by R.C. Read, 'An Introduction to Chromatic Polynomials," *Journal of Combinatorial Theory*, Vol. 4, 1968, 52-71.

5. The two greedy algorithms for the minimum spanning tree problem, method 1 and method 2, were presented by Prim and Kruskal, respectively. See R.C. Prim, "Shortest Connection Networks and Some Generalizations," *The Bell System Technical Journal*, Vol. 36, 1957, 1389-1401. See also J.B. Kruskal, Jr., 'On the Shortest Spanning Subtree of a Graph and the Traveling Salesman Problem," *Proceedings of the American Mathematical Society*, Vol. 7, 1956, 48-50.

6. The algorithm in this section uses a simple labelling technique developed by U. Pape and D. Conradt, 'Maximales Matching in Graphen," in *Ausgewählte Operations Research Software in FORTRAN*, edited by H. Späth, Oldenburg, Munich, 1979, 103-114, for implementing the maximum matching algorithm originally given by J. Edmonds, "Paths, Trees and Flowers," *Canadian Journal of Mathematics*, Vol. 17, 1965, 449-467.

7. The linear time planarity testing algorithm is the contribution of J.E. Hopcroft and R.E. Tarjan, "Efficient Planarity Testing," *Journal of the ACM*, Vol. 21, October 1974, 549-568.

INDEX